BIOINSECURITIES

ANIMA

A series edited by Mel Y. Chen and Jasbir K. Puar

BIOINSECURITIES

Disease Interventions, Empire, and the Government of Species

NEEL AHUJA

DUKE UNIVERSITY PRESS *Durham and London* 2016

Printed and bound by CPI Group (UK) Ltd, Croydon, CR0 4YY
Typeset in Minion by Tseng Information Systems, Inc.

Library of Congress Cataloging-in-Publication Data
Names: Ahuja, Neel, [date] author.
Title: Bioinsecurities : disease interventions, empire, and the
government of species / Neel Ahuja.
Other titles: ANIMA (Duke University Press)
Description: Durham : Duke University Press, 2016. |
Series: ANIMA | Includes bibliographical references and index.
Identifiers: LCCN 2015038468
ISBN 9780822360483 (hardcover : alk. paper)
ISBN 9780822360636 (pbk. : alk. paper)
ISBN 9780822374671 (e-book)
Subjects: LCSH: Imperialism—Health aspects. | Medical
anthropology. | Biosecurity. | International relations. |
Biopolitics. | Diseases in literature. | Health in literature.
Classification: LCC R133 .A385 2016 | DDC 610.9—dc23
LC record available at http://lccn.loc.gov/2015038468

Cover art: *A Misogynist Monkey Seeks Solitude in the
Caribbean off Puerto Rico*, 1939. Photograph by Hansel
Mieth. The LIFE Picture Collection / Getty Images.

BIOINSECURITIES

Disease Interventions, Empire, and the Government of Species

NEEL AHUJA

DUKE UNIVERSITY PRESS *Durham and London* 2016

© 2016 Duke University Press
All rights reserved
Printed and bound by CPI Group (UK) Ltd, Croydon, CR0 4YY
Typeset in Minion by Tseng Information Systems, Inc.

Library of Congress Cataloging-in-Publication Data
Names: Ahuja, Neel, [date] author.
Title: Bioinsecurities : disease interventions, empire, and the
government of species / Neel Ahuja.
Other titles: ANIMA (Duke University Press)
Description: Durham : Duke University Press, 2016. |
Series: ANIMA | Includes bibliographical references and index.
Identifiers: LCCN 2015038468
ISBN 9780822360483 (hardcover : alk. paper)
ISBN 9780822360636 (pbk. : alk. paper)
ISBN 9780822374671 (e-book)
Subjects: LCSH: Imperialism—Health aspects. | Medical
anthropology. | Biosecurity. | International relations. |
Biopolitics. | Diseases in literature. | Health in literature.
Classification: LCC R133 .A385 2016 | DDC 610.9—dc23
LC record available at http://lccn.loc.gov/2015038468

Cover art: *A Misogynist Monkey Seeks Solitude in the
Caribbean off Puerto Rico*, 1939. Photograph by Hansel
Mieth. The LIFE Picture Collection / Getty Images.

CONTENTS

Even at their most powerful, states are rarely ever able to remake life in their own sovereign image.[1] The difference in the case of an imperial state is not simply that it has incrementally more power to control, but that its scale of intervention and fantasies of mastery doom it to ever more desperate interventions that seek to postpone its overextension and demise. Imperial states extend tentacles of intervention into varied domains of life in order to displace the crises of reproduction and legitimacy they inevitably generate.

Given this precarious, adaptive, and expansionist form of empire, it is unsurprising in the early twenty-first century that basic controversies about the protection of biological life have become major concerns of a United States that brands its own rule as a defense of democracy and thus both freedom and life. Indeed, as in past eras of empire, many of the most intimate bodily experiences of living and dying drive political debate.[2] With steady media attention to health care policy, humanitarian intervention, abortion, prenatal genetic testing, drug patents, assisted suicide, animal rights, environmental regulation, and biosecurity, the idea of the vulnerable body as an object of governance—its transformation for better or worse through state policies, technological intervention, and ecological forces—is today quite conventional.

One element common to this biopolitics of empire is an anxiety about the dependence of the human body on forces that appear inhuman, even inhumane: medical technologies to extend, optimize, or end life; markets and institutions that unequally distribute resources for sustaining life; environmental processes that support, deprive, or injure bodies. Such concerns were, of course, entirely common to twentieth-century modernist fears of alienation from nature, as well as to liberal, socialist, and fascist states that each proclaimed to defend the life of the people in the major imperial wars. Yet due to the ongoing expansion of government into life through technological, economic, and environmental interventions, a growing number of crises that advertise dreaded risks to life as we know it—climate change, nuclear toxicity, disease pandemics, biological weapons, and financial speculation, to name a few—have recently pressed critical studies of empire to think politics and agency at queer scales of relation, from the grand vantage of planetary geology and climate, through the lively migrations of commodities and animals, all the way down to the microbial, molecular, and quantum worlds of matter in which advanced sciences produce new technologies and knowledge. In an era in which excessive hope is invested in the idea that empire's so-called free markets will inevitably deliver resources for improving life, discussions of risk and security increasingly provoke concern about how bodies are either threatened or safeguarded in links to other species, to ecology, and to technology. Public fears and hopes are thus invested in questions about how bodies interface beyond the skin of the organism. The living body is not only an ecology reproduced by constituent species (think of the life-sustaining work of gut bacteria or the ingested flesh of animals or plants). It is also an assemblage crosscut by technological, economic, and environmental forces (medical technologies, insurance markets, agricultural systems, toxic pollution) that render the body vulnerable as they reproduce its conditions of possibility.

Yet there remains a sense of tension concerning how social theorists frame the vulnerability of human life between biopolitics and these emerging posthumanist ideas. While biopolitical analysis foregrounds the contested figure of the human, emphasizing that the human body is an effect of power crafted through the social reproduction of nationality, race, sex, and/or class factors conjoined in inhuman fields of power, emerging posthumanist and new-materialist fields including animal studies, environmental humanities, and object-oriented ontology more often emphasize the agency of the nonhuman and the surprising liveliness of physical matter. As such, despite the avowed critique of the human, they may take for granted the apparent universality of the human lifeworld from which they flee, foreclosing attention to the processes that anthropomorphize the human in order to characterize the human's sovereign domination of the nonhuman. This move allows some posthumanist critics to project upon an outside, the nonhuman (in the form of environment, animal, machine, or other object), the possibility of resistance to anthropocentrism. Such thinking might be seen as a ruse of transcendence—an assumption that turning attention from the human to the nonhuman could bypass Marxist, feminist, critical race, and postcolonial critiques of imperial systems that proliferate inequality under the guise of universal human freedom.

Despite this liberal, idealist trend among posthumanists (which is more pronounced in the humanities than it is in the social sciences), studies of empire increasingly confront the fact that the apparent exteriority of the subject (the worlds of body, physical matter, and interspecies exchange) has more often formed the center of the politics of empire rather than its excluded outside. It is thus my hope that the collision of biopolitical and posthumanist thought may be salvaged in a practical if unexpected crossing: a more robust accounting of the ways in which politics, including the liberal and neoliberal politics of empire, is embedded in living bodies and planetary environments, which are themselves constituted as objects of knowledge and intervention for imperial science. Such an understanding goes beyond an assertion that life is controlled by human government, which would embrace the strong post-Enlightenment division between government and life, human and nonhuman. I instead hope to explore the queer hypothesis that the adaptability, risk, and differentiation central to life increasingly constitute the very matter of politics. This book is about how disease outbreaks, medical technologies, and the relations between humans, animals, bacteria, and viruses galvanized racialized fears and hopes that determined the geopolitical form of US empire during the long twentieth century, following the continent-wide establishment of Euro-American settler networks. Before explaining that argument, however, this brief preface explores how—in addition to established methods of postcolonial study that define empire through histories of conquest, settlement, and the exploitation of labor and resources—the inequalities and violences of imperialism can productively be understood from the vantage of species, the field of life itself.

Research on colonial environmental history and disease control is long established in postcolonial studies, even as today there is growing attention to Global South environmental activism, advanced biotechnologies, and human-animal and human-plant interactions as significant concerns in the planetary routes of European and US empire. Yet my sense of an interspecies politics is still relatively unfamiliar from even the vantage of these studies. Extant studies have long highlighted questions of representation, agency, influence, and domination, explaining the unequal distribution of the privileges accorded for being anthropomorphized, for being made human through colonial ideological and social processes. While maintaining focus on such racialized inequalities fracturing the figure of the human in the worldwide routes of European and US imperialisms, it is the aim of this book to articulate an additional sense of the political as a lively zone of embodied connection and friction. "Interspecies relations form the often unmarked basis upon which scholarly inquiry organizes its objects, political interventions such as 'human rights' stake their claims, and capitalist endeavors maneuver resources and marshal profit."³ A critique of the interspecies zone of the political—which at its broadest would expand beyond the human-animal and human-microbial relations discussed in this book to include the diversity of living species, matter, energies, and environmental systems that produce everyday life out of biosocial crossings—helps us understand the persistence of empire in a postcolonial age precisely because it conjoins power to forces that retreat into the seemingly natural and ahistorical domains of body and matter. From this vantage, empire appears not only as a process of territorial and economic accumulation across international divisions of labor and sovereignty, but also as a reproductive process managing bodies in unequal planetary conjunctions of life and death. Tracing this second phenomenon requires analysis of biosocial forms of exchange among microbes, plants, animals, and humans, as well as models of power and representation recognizing that bodies are not empty containers of human political subjects, but are lively, transitional assemblages of political matter.⁴

There are risks in attempting to theorize a political process like empire via the material shape it takes in life and matter, anticipated in long-standing liberal and Marxist distinctions between human and natural history.⁵ Must such a move necessarily turn away from issues of interest, hegemony, violence, representation, and inequality that often define organized decolonial struggles? I would argue that this need not be so, and that vitalizing colonial discourse studies through an accounting of empire's living textures may actually give a more grounded account of imperial power as well as the strategies of representation that have persistently masked its material articulations.⁶ To this end, I explore empire as a project in the government of species. Broadly, this idea refers to how interspecies relations and the public hopes and fears they generate shape the living form and affective lineaments of settler societies, in the process determining the possibilities and foreclosures of political life. In practice, the government of species has historically optimized and expanded some life forms (human or otherwise) due to biocapital investments in national, racial, class, and sex factors. Operating through interspecies assemblages known as bodies, such investments selectively modify and reproduce life forms and forms of life, extracting "the human" out of the planetary field of interspecies relation. Once securitized, this form is constantly under pressure from the unpredictable and inhuman risks of life in a world of ecological, economic, and political complexity. These forces in turn contribute to the ways publics experience and interpret their futures as more or less livable.

An account of the government of species thus explains that empire can be understood as a project in the management of affective relations—embodied forms of communication and sensation that may occur independently of or in tandem with sentient forms of thought and discourse. These affective relations cross the divisions of life and death, human and animal, media and bodies, and immune and environmental systems. In the process of forming the human out of cacophonous biosocial relations, empire often persists—even after the formal conclusion of colonial occupation or settlement—in part because it invests public hope in the management of bodily vulnerability and orients reproductive futures against horizons of impending risk, a phenomenon I call *dread life*. In such processes by which bodily vulnerability is transmuted into political urgency, techniques proliferate for managing the relations of populations and the living structures of species (human, animal, viral). As such, empire involves the control of life through accumulation of territory and capital, which may be securitized by activating life's relational potential. Lauren Berlant describes a "lateral agency" that moves across bodies and populations rather than in the top-down fashion of sovereign power; it may, then, be possible to understand empire's force of securitization not only through conventional dramas of domination and resistance, but rather through embodied processes of coasting, differentiating, adapting, withering, transition, and movement.[7] These are processes that subtly determine how bodies take form, and to what extent they are able to reproduce themselves in space-time relation. They also more radically stretch the body beyond the organic lifetime and into evolutionary, environmental, and informational domains where life/death distinctions blur.

However, the intimate connection between the governmental imperatives to make live and to make die, which Jasbir Puar names "the bio-necro collaboration," has long been obscured in social and political theories.[8] It thus remains commonplace for biopolitical analyses to view power as either repressive or productive in essence. In his classic work on the topic, French philosopher and social theorist Michel Foucault argued that by the eighteenth century, a political form had emerged in Europe targeting the human as biological species as the central object of power. Power was no longer simply about the repressive force of the state and its controlling interests wielding the right to kill. Power was increasingly vested in the productive reshaping of the biological life of human organisms by institutions such as clinics, prisons, and asylums and their related forms of scientific knowledge; power meant letting live, albeit in constrained form.[9] Foucault recognized the embedding of biopower across species, calling for a social history that incorporated "the evolution of relations between humanity, the bacillary or viral field, and the interventions of hygiene, medicine, and the different therapeutic techniques." In the notes to his late lectures, he even speculated that neoliberalism involved a governmentality that can "act on the environment and systematically modify its variables."[10]

Foucault's description of the rise of biopower is the inspiration for a number of studies in sociology and anthropology that assess new biopolitical shifts involving advanced biomedical technologies.[11] Given that these biopolitical studies focus largely on the United States, western Europe, China, and India—states that have built biotechnology sectors as engines of unequal neoliberal growth—it is perhaps not surprising that a concomitant line of critique has emerged acknowledging vast and growing world sectors of biological and economic precarity. Building on a number of key postcolonial/feminist studies of the 1990s exploring Foucault's theory beyond European borders, these necropolitical critiques announce that politics today often emerges as the specter of death.[12] The world's poor, as well as a growing "precariat" carved from shrinking national bourgeoisies, appear less often as the objects of technological uplift than as the human surplus of the political order of things, populations at risk for displacement, dispossession, captivity, and premature death. The precaritization of sweated labor, the subjection of agrarian populations to the twin scourges of neoliberal structural adjustment and environmental devastation, the proliferation of deterritorialized war and ethnic cleansing, and the growth of predatory industries and rents to recycle capital from surplus populations all reveal that those humans targeted for biopolitical optimization constitute a shrinking population who reproduce through the cannibalistic appropriation of life elsewhere. But necropower is not simply about the distribution of death; it is also about the accumulation of social or economic capital through death and precarity. For example, when suicide passes on social force through the deathly body, or when life insurance capitalizes death, death itself thus gives form to life.

The most compelling of these studies undermine normative divisions between life and death. In the process, these two emphases within biopolitical analysis—productive forces marshaled through remaking life and deploying death—have invigorated critiques of neoliberal globalization, tracking the transnational shaping of the human by the inhuman forces of sovereignty and capital. What would it mean, then, to think these bio-necro coordinates as materialized in interspecies processes of empire building? What would it mean to assert not simply that life, death, and politics have become inextricably linked, nor that life is an effect of politics, but that life gives form to the political? This question—of how power could be materialized through processes of interspecies exchange and affective transit that work through yet leak beyond structural organization of human agency and cognition—is a question that Foucault never fully articulated in his late studies following the description of biopower in the first volume of *The History of Sexuality*. Disappointed in the late 1970s by, on the one hand, the failures of the Iranian and Vietnamese revolutions to provide a challenge to US empire, and, on the other, the rise of neoliberal free trade agendas, Foucault famously made an ethical turn and sought radical modes of living in the histories of ancient Greece and early Christian Europe. In his reporting on Iran, he even embraced a "political spirituality" idealizing religious ethical conducts occluded by the rise of liberal secularism.[13] If biopower was a description of how the human was produced through inhuman biosocial forces, it is possible that Foucault's turn to masculinist visions of pastoral and spiritual life in these late writings reflected a nostalgia for a time before biopower had been captured by the conjoined forces of neoliberal capital and US imperial militarism that appeared increasingly dominant at the end of the Cold War. There has been much debate since on whether Foucault in the process abandoned his earlier critiques of the subject in the name of an individuated (even humanist) sense of ethical self-constitution. Yet I wish to pursue a different set of questions, exploring in this book whether Foucault's theorization of biopower could have been extended to explain how empire's racial government of the human as species approached the interspecies horizons of planetary life and death.[14] This preface suggests that we might, as an alternative to Foucault's or the posthumanists' attempts to transcend politics through ethics, draw out biopolitical critique into the material arenas of life and planet that empire takes as its space of intervention.[15]

What understandings of the political, of embodiment, of state, of war, and of life itself emerge by tracking animal and viral transits through the splintered figure of the human? Postcolonial theory has contributed to a critique of the human by tracing empire's compression of life into raced and gendered figures of self-possession. Lisa Lowe thus characterizes the liberal human subject as an empty formalism abstracted from transcolonial intimacies of slavery, indenture, and revolution.[16] It may, then, be possible to track the intimate relations, violences, and exchanges that emanate from interspecies projects to reproduce colonial forms of property in life against the tandem proliferation of dispossession, captivity, and waste. Such work follows on a history of work on nature, embodiment, technology, affect, and animals within feminist science studies, which guides my critiques of posthumanist and new-materialist metanarratives.[17] This work has developed in tandem with animal and environmental movements since the 1970s, as well as with the rapid expansion of new biotechnologies. Yet more work remains to be done to explore the crossings of feminist and postcolonial perspectives on such phenomena. Little has been written on the colonial genealogies of the posthumanist turn. The racialized form of posthumanist knowledge projects is particularly evident when environments or animals are rendered through tropes of wilderness external to the human or when turns to animals, environments, and things rely on a figure of unmarked whiteness in the form of the universal human. Such notions may dovetail with the efforts of the racial security state to conceal its forms of containment aimed against "surplus populations" whose bodies and homes are rendered as sites for the dumping of risk.[18] Furthermore, posthumanism evolves through state and capital forces that turn to the plasticity of life to find what Melinda Cooper calls "vital fixes" for the reconfiguration of war and the recycling of surplus into capital. Biological life itself—down to the informational bodies of DNA—is being remade as a privileged locus of property and surplus.[19] Contrary to idealist descriptions of hybridity, difference, and queerness as inherently liberatory, Nicole Shukin claims that we "inhabit an anthropocentric order of capitalism whose means and effects can be all too posthuman, that is, one that ideologically grants and materially invests in a world in which species boundaries can be radically crossed."[20] Moving beyond Foucault's conception of a biopower that constituted the human as species, today empire technologizes the mutable borders of species as one horizon of accumulation.

The mobilization of fear and anxiety concerning "the outside" of the human subject emerges from Euro-American empire's history of racial engulfment.[21] The post-Enlightenment fashioning of the human always incorporated a racialized indeterminacy which posited that feminized and colonized bodies—enmeshed in the inhuman worlds of nature—were particularly affectable, were defined by their physical exteriority rather than interiorized by the human subject's universal capacities for reason, language, and sentiment.[22] Yet at the same time, the practical administration of colonial settlement has been preoccupied by the mundane problems of reproducing the settler bodies through which this universal human political subject could emerge. Thus contemporary tropes of networks, systems, complexity, assemblage, and vitality that work through the affectable matter of bodies carry ambivalent traces of colonial subject and settlement fashioning that extracted the figure of the human from immanent ecologies of transcolonial production and consumption. One effect of this situation is a risky intimacy of post-humanist and new-materialist critical tropes with exoticist and orientalist attempts to turn from the Cartesian subject to insurgent matter, from sovereign human to subaltern nature. Donna Haraway once insightfully called the field of primatology a form of "simian orientalism" for this very reason: in the speculative engineering of primate bodies we may find not just a corrective to empire's compression of life into the immaterial container of the human, but also a scene of universal origins and a material for the projection and modeling of racial and sexual difference.[23] Twenty-five years after its publication, Haraway's magnum opus, *Primate Visions*, remains the only book-length postcolonial critical account of the emergence of systems theory. Since its publication, the posthuman turn and, more specifically, the metaphorics of systems and complexity have become central to critiques of anthropocentric distinctions between society and environment, but these trends have largely obscured Haraway's insistence on systems theories' enmeshment in histories of colonial warfare and racialization. The systems-theory tradition, extending from the cybernetics of Ludwig von Bertalanffy and the sociology of Talcott Parsons and Gregory Bateson to the neostructuralism of Niklas Luhmann and the biomysticism of Humberto Maturana and Francisco Varela, appears to oppose complexity and holism to various forms of scientific and social reductionism; however, the tropes of the system-environment relation oddly introduce their own reductionism by systematizing everything, putting everything into relation despite the possibilities of segregation, expulsion, individuation, or dimensional phase shifts which this book argues are significant elements of bio-necro collaboration. Settler-colonial conditions furthermore enabled thinking ecologically as antidote to high-modernist rationality. For example, early systems theorist Talcott Parsons's statement of American exceptionalism emphasizes the apparently unique social, psychic, and ecological complexity of Puritan networking in the colonial settlement of North America, an argument that oddly complements Gilles Deleuze and Félix Guattari's orientalist romance of American Indians as privileged site of rhizomatic becoming.[24]

Romantic notions of the otherness of environments and species are entirely compatible with forms of government aimed at imperial containment. Post-humanist knowledges plainly serve as resource for both the agents of empire and its critics. How, then, might we disentangle posthuman idealism from the government of species? Is there a way of breaking through anthropocentric methods for Left political critique without dismissing the imperial genealogies of posthumanism or the racialized violence of its incorporation into capital? How can critical theory offer a methodological generosity toward life and matter without romancing bodies, animals, microbes, cyborgs, and things, disavowing their assimilation into international divisions of labor and life? To begin to answer such questions, the introduction and chapters that follow explore how US empire has long been constituted as a project crossing borders and species, a project that embeds the national defense and imperial interest in the racialized matter that we call bodies. In the process, they give an account of how the value-added body of the settler emerges from a form of empire that increasingly takes on the planetary field of life itself as its space of intervention.

ACKNOWLEDGMENTS

Thanks to my teachers. Michael Davidson and Page duBois, my mentors from the Department of Literature at the University of California–San Diego, have been consistent supporters of this book. I am inspired by them and by other feminist thinkers whose writings have taught me to envision politics beyond the fear of death and the skin of the human: Octavia Butler, Lauren Berlant, Mel Chen, Colin Dayan, Jack Halberstam, Donna Haraway, Lisa Lowe, Renisa Mawani, Jasbir Puar, Denise Ferreira da Silva, and the late Rosemary Marangoly George. Special thanks to Cindy Burgett, Diana Davis, Ji-Yeon Yuh, and Jillana Enteen, who nurtured my interests in literature and politics when I was a high school student in Topeka, Kansas, and an undergraduate at Northwestern University. Everything I know about power I learned from fellow activists working on international labor, animal, and antiwar campaigns; during my time in Chicago, Jenny Abrahamian, Desiree Evans, Mischa Gaus, Dana Lossia, Pete Micek, Chris Sherman, Jake Werner, and many others taught me through shared struggle. My interests in health, medicine, and migration have also been deeply shaped by the stories of my father, Sain Ahuja, and his experiences as both a medical physicist and a refugee of the war of Indian partition in 1947.

I could not have written this book without learning from brilliant graduate and undergraduate students including Aisha Anwar, Nicole Berland, Laura Broom, Patrick Dowd, Adam Faircloth, Wayland Ferrell, Tegan George, Amanda al-Raba'a, Kriti Sharma, Pavithra Vasudevan, Peter Warrington, and L. Lamar Wilson. I also owe thanks to the many faculty colleagues who gave me encouragement, feedback, or other support during the writing process: Aimee Bahng, Colleen Boggs, Michael Davidson, Zulema Diaz, Page duBois, Steven Epstein, Rebecka Fisher, Carla Freccero, Rosemary George, Minrose Gwin, Jennifer Ho, Julietta Hua, Alexandra Isfahani-Hammond, Jinah Kim, Michael Lundblad, Tim Marr, John McGowan, Chandan Reddy, Peter Redfield, Jamie Rosenthal, Ruth Salvaggio, Kyla Schuller, Nayan Shah, Nitasha Sharma, Denise Ferreira da Silva, Kathryn Shevelow, Elizabeth Steeby,

Shelley Streeby, Matt Taylor, Jane Thrailkill, Priscilla Wald, Winnie Woodhull, Kathleen Woodward, and Ji-Yeon Yuh. I received feedback from faculty, students, librarians, and independent scholars following presentations of parts of this work at the University of British Columbia, the University of California–Davis, the University of California–San Diego, the University of California–Santa Barbara, the University of Chicago, Duke University, the University of North Carolina, Yale University, and the meetings of the Association for Asian American Studies, the Society for Caribbean Studies, and the American Studies Association. The revision of the manuscript benefited from my participation in a seminar on Sex, Gender, and Species facilitated by Ranjana Khanna and Kathy Rudy in the Women's Studies Department at Duke University and from a series of events organized by María Elena Garcia and Louisa Mackenzie of the Postcolonial Animal Working Group at the University of Washington.

My research was funded by the University of California President's Dissertation Fellowship Program, California Cultures in Comparative Perspective, UCSD Center for the Humanities, the Junior Faculty Development Award Program of the UNC Provost, the UNC University Research Council, and the UNC Department of English and Comparative Literature, where department chair Beverly Taylor was instrumental in helping me secure writing time and resources. I relied heavily on the staffs at the University of California Libraries, the University of North Carolina Libraries, the Library of Congress Newspapers and Periodicals section, the India Office Records at the British Library, the Rubenstein Rare Book and Manuscript Library at Duke University, and the History of Medicine Reading Room at the National Library of Medicine. Chapter 3 would not have been possible without the input and assistance of Claud Bramblett, formerly of the Darjani primate station, Kenya; Gabriel Troche, formerly of Cayo Santiago station, Puerto Rico; George Siebert, film collector; and the staff at Kensington Video, San Diego. Thanks to D. A. Henderson and John Wickett for assistance with an image for chapter 4.

Early drafts of sections of chapters 1 and 3 appeared in a different form as "The Contradictions of Colonial Dependency: Jack London, Leprosy, and Hawaiian Annexation," *Journal of Literary and Cultural Disability Studies* 1, no. 2 (2007): 15–28; "Macaques and Biomedical Discourse: Notes on Decolonization, Polio, and the Changing Representations of Indian Rhesus in the United States, 1930–1960," in *The Macaque Connection: Cooperation and Conflict between Humans and Macaques*, edited by Sindhu Radhakrishnan, Michael A. Huffman, and Anindya Sinha (New York: Springer, 2013), 71–91; and "Notes on Medicine, Culture, and the History of Imported Monkeys in Puerto Rico," in *Centering Animals in Latin American History*, edited by Martha Few and Zeb Tortorici (Durham, NC: Duke University Press, 2013), 180–205.

I am especially thankful for the assistance of Duke University Press in supporting me as a first-time author. Series editors Jasbir Puar and Mel Chen were supportive of the project from the beginning. Courtney Berger was an incisive and supportive editor who made the revision process both rigorous and humane. I am grateful for the generous and patient feedback of Colleen Boggs and one other anonymous manuscript reviewer. Their meticulous comments and multiple readings of the complete manuscript were pivotal in sharpening the arguments, structure, and critical voice of the book. Erin Hanas, Liz Smith, and Christine Riggio made important contributions during the production process.

I acknowledge that this book was written on lands traditionally occupied and transited by the Kumeyaay, Potawatami, Miami, Eno, Shakori, and Sissipahaw peoples of California, Illinois, and North Carolina. It is my hope that it may in some small way contribute to decolonial struggles in North America and beyond.

Finally, I must thank those kin—human, canine, and feline; living in the United States and India; related by blood, friendship, pleasure, cohabitation, and political community—who offered me space, food, laughter, affection, resources, and patience during the ten years I spent writing. Special thanks to my entire extended family—especially to Parmod Dewan, Mamta Dewan, Sheenu Dania, Rano Ummat, Tarun Ummat, Harsween Ummat, and Anand Verma, who helped me pursue archival research and language study in India. I am grateful for the love and support of Jamie Rosenthal, Chef, Keiki, Jenny Abrahamian, Usha Ahuja, Sain Ahuja, Sonia Ahuja, Shyla Ahuja, Sarita Ahuja, Andras Ambrus, Dana Lossia, Jinah Kim, Peter Holderness, Chris Sherman, Neha Bhardwaj, Meghann Wilkinson, Julietta Hua, Aimee Bahng, Chuong-Dai Vo, Kyla Schuller, Elizabeth Steeby, Denise Khor, Sara Forrest, Brenda Lehman, Cheryl Taruc, Jake Peters, Zulema Diaz, Rohan Radhakrishna, Nina Martin, Jean Dennison, Mike Ritter, Sara Smith, Tonyot Stanzin, Erika Wise, Aaron Beyerlin, Jocelyn Chua, Jennifer Ho, Matthew Grady, Ariana Vigil, Laura Halperin, Heidi Kim, Jes Boon, Sarah Bloesch, Hong-An Truong, and Dwayne Dixon. As this book went into press, I welcomed my daughter and comrade Naya to the world. And as life somehow persists in this age of extinction, I think finally of elders who nurtured me across oceanic expanses: this book is dedicated to the memory of my grandfather, Prem Nath Dewan.

Dread Life

Disease Interventions and the Intimacies of Empire

In every well-ordered society . . . the rights of the individual in respect of his liberty may at times, under the pressure of great dangers, be subjected to such restraint, to be enforced by reasonable regulations, as the safety of the general public may demand. An American citizen arriving at an American port on a vessel in which, during the voyage, there had been cases of yellow fever or Asiatic cholera, he, although apparently free from disease himself, may yet, in some circumstances, be held in quarantine against his will on board of such vessel or in a quarantine station. . . . The liberty secured by the 14th Amendment, this court has said, consists, in part, in the right of a person "to live and work where he will"; and yet he may be compelled, by force if need be, against his will and without regard to his personal wishes or his pecuniary interests, or even his religious or political convictions, to take his place in the ranks of the army of his country, and risk the chance of being shot down in its defense.
—JUSTICE JOHN MARSHALL HARLAN, *Jacobson v. Massachusetts*, February 20, 1905

A smallpox epidemic is so vicious and kills so many people so rapidly and spreads far and wide, that, after a great deal of thought, I concluded that the US military people who have potential vulnerability ought to [be vaccinated].
—DONALD RUMSFELD, interview with Larry King, December 18, 2002

How does the concept of national defense materialize in the form of living bodies? I begin with two events that serve as entry points for a theory of the body as a transitional theater of imperial warfare. On July 17, 1902, during the last smallpox epidemic to hit Boston, a Swedish immigrant named Henning Jacobson refused an order by the Board of Health of Cambridge for his son and himself to receive the city's mandatory vaccinations. He was accordingly fined five dollars (the equivalent of one hundred US dollars in today's currency). Jacobson sued, contending that the state's requirement violated his personal liberty and that vaccination was an unsafe practice. In 1905, the US Supreme Court ruled in *Jacobson v. Massachusetts* that a state does have the right to enforce compulsory vaccination, quarantine, or other public health protections. The majority opinion makes disease control a national defense priority: "Upon the principle of self-defense, of paramount necessity, a community has the right to protect itself against an epidemic of disease which threatens the safety of its members." As the phrasing of the decision compares bacterial and viral species to military enemies, it invokes the specter of disease as cause for emergency intervention. According to Justice John Marshall Harlan's majority opinion, public health authorities acted "under the pressure of great dangers" and should thus be afforded police powers to impose emergency solutions—including ones that may harm individuals. This sacrifice they take on echoes the drafted soldier's "risk . . . of being shot down in [his country's] defense."

Save for the lawsuit that brought his story to national prominence, Jacobson's refusal of vaccine may have appeared an unremarkable incident in the history of US public health. Liberal narratives of biomedical progress assert that health and medical innovations inevitably make humans and their environments more safe, healthy, and sanitized, despite the anxieties of skeptics who question modern health expertise. Yet Jacobson and other "conscientious objectors" to vaccination had a rather significant following at the time. They considered the cow-derived smallpox serum unsafe, unsanitary, and potentially lethal. Controversies over vaccination, vivisection, quarantine, and other biopolitical interventions were hotly debated across the US and British empires, from India to the US-Mexico borderlands. Such conflicts emerged at a time when US officials were increasingly concerned with the uncertainties of exposure to distant lands and bodies, especially through contact with a rising Asia.[1] It is thus no wonder that later passages of the *Jacobson* decision take pains to detail the widespread use of enforced vaccination in both the British Isles and colonial India. The court saw transborder outbreaks as an appropriate arena of governance for empire-driven states.

Thus the consequences of Henning Jacobson's refusal of smallpox vaccine reverberated far beyond Cambridge, Massachusetts. They involved concerns that contagions would pass through the bodies of travelers, immigrants, traders, animals, soldiers, sex workers, and the colonized, whose planetary routes of movement and contact suggested that the biological character of the nation was being remade via expansion. At the same time, the *Jacobson* precedent (which remains standing law) underpinned expanding forms of social control in the twentieth century, including the court's infamous ruling up-holding forced sterilization.[2] After the long warfare of European settler ex-pansion across the continent and the genocidal displacement of American Indian nations, the transition from a continental to a planetary empire invigo-rated concerns about disease that would buttress new emergency state powers. Notably, the "American citizen" to be drafted in defense of public health was one who crossed continents; the Court figures disease-causing epidemics as originating from outside of the continental borders. Justice Harlan, who in a separate case attempted to deny birthright citizenship to the children of Asian immigrants, posed state police power as the solution to smallpox, "Asiatic" cholera, and yellow fever — contagions that had long been associated with im-migration, urbanization, and the western and southern frontiers of Anglo-American settler colonialism. In this sense, the *Jacobson* decision confirmed that standard public health interventions were not simply public goods, but were indeed the privileged state avenue for defending the national body in a world of expanding contact.

One hundred years after Jacobson's refusal of vaccine, US secretary of de-fense Donald Rumsfeld rearticulated Harlan's ethic of national sacrifice in de-fense against smallpox. In late 2002, Rumsfeld, along with Vice President Dick Cheney, promoted a campaign to convince the US military, first responders, and the general public of the need for smallpox vaccination. The two warned that Iraqi president Saddam Hussein — whom they depicted as a rogue secretly allied with al-Qaeda — had the wherewithal to weaponize variola virus. They thus promoted what was known to be an unusually risky live smallpox vaccine.[3] Rumsfeld advertised in a television news interview that, along with President George W. Bush, he would take the vaccine himself in order to demonstrate its safety and efficacy: "I certainly intend to [take the vaccine], simply because it's hard to ask people to do something that you're not willing to do yourself."

The smallpox program was just one of a host of new biosecurity measures that were adopted following the September 11, 2001, attacks. These included establishment of a network of port quarantines at twenty-five US points of entry and three new multibillion-dollar programs — BioSense, BioShield, and BioWatch — to centralize data on potential disease outbreaks, to increase drug development capacity and vaccine stockpiles, and to monitor the envi-ronments of major population centers. Yet the smallpox program was both the most high-profile and the most controversial of these projects. Inverting

Jacobson's refusal of vaccine during an epidemic, Rumsfeld volunteered to take vaccine in the absence of even a single case of infection. Rumsfeld portrayed smallpox, like Saddam himself, as "vicious," claiming the virus "kills so many people so rapidly and spreads far and wide" that it risked catastrophic harm.[4]

This fearful rendering of smallpox as a nefarious, intentional enemy ran up against a complication: smallpox did not exist in the environment. The variola virus had been eradicated in a coordinated global public health campaign in the 1970s. If the Boston smallpox epidemic of 1901–3 made transparent to the Supreme Court the "great danger" requiring health policing powers, Rumsfeld's move scrambled the relationship between public signals of harm, state intervention, and biological risk as the administration attempted to convince a skeptical public to go to war in Iraq. In this case, vaccination was the signal of smallpox risk rather than its solution; the prick of the needle could localize the otherwise diffuse sensation of the soldier's vulnerability to bioweapons. The optimization of Rumsfeld's immune system in US media turned a technique of public health into a technique of public relations, signaling that American bodies and vital systems (such as hospitals and the military) were vulnerable to the designs of faraway enemies. As demonstrated more recently with the CIA's use of a vaccination campaign to locate Osama bin Laden in Pakistan, vaccination could be transformed from a medical defense into a weapon.[5] In 1902 a known viral epidemic demanded the sacrifice of individual liberty for the cause of vaccination. In 2002, vaccination made a spectacle of physical vulnerability that could be mobilized for the cause of war itself.

Empire of Risk

These two events — Jacobson's challenge to the Cambridge Board of Health and Rumsfeld's launch of the Smallpox Vaccination Program — serve as bookends for my investigation of US empire as a project to govern environments and the material form of bodies, human and nonhuman. *Bioinsecurities: Disease Interventions, Empire, and the Government of Species* explores how disease control proliferated the expansion of US empire into the domain of biological life over the long twentieth century, 1870–present.[6] Examining projects targeting smallpox, syphilis, polio, Hansen's disease (leprosy), HIV, and other infectious diseases, *Bioinsecurities* contends that the form of US empire has been deeply conditioned by entanglements between microbial species, nonhuman animals, and unequal human populations navigating environments of military, economic, and territorial expansion. Fearing engulfment in the geographies of disease transmitted through the bodies of travelers, immigrants, traders, animals, soldiers, sex workers, and the colonized, the forces of imperial disease intervention constituted settler bodies and ecologies as an emergent space of technocratic control, rendering them lively domains of warfare. As depicted in the statements of Harlan and Rumsfeld, these interventions operate through a number of logics, including attempts to prevent the geographic transmission of disease across borders and species, the use of drugs including vaccines to prevent infection, the development of large-scale systems of surveillance and quarantine, and efforts to control how the public feeling of risk circulates between experts and broader publics.

Even as the work to control disease incorporated multiracial populations into the "imagined immunities" of the settler nation, the purported universality of imperial public health was betrayed by its circulation of racial fears of disease.[7] This made the microscopic bodies of viruses and bacteria into the very matter of racial differentiation, the lively conduits of debility and death that threatened a dangerous intimacy between species and social groups in a globalizing world of empire. In response, disease interventions at the borders of expanding US influence deployed models of territorial warfare to defend the body in space, but also transformed the body's actual biological processes into a site of management and optimization using advanced biosciences and animal research subjects. In the process, a physics of imperial power emerged conjoining repressive control of bodies to attempts to add capacity to the body's immune environments. This merger of repressive control with humanitarian health and animal biocapital was significant given that the twentieth-century rise of working-class and decolonial movements required European and US governments to disavow empire and disguise wars and other forms of state intervention as expansions of liberal freedom. This book, then, describes a conjunction of US empire with an embodied politics of race and species. Such an analysis requires taking seriously the materiality of bodies, whose surfaces and internal structures can be surveyed and managed through forms of scientific knowledge and activation that are fluid and adaptive to the contexts of politics, economy, and warfare.

More specifically, this book argues that the racialization of transborder epidemics—the use of media to activate the feeling of bodily risk through the touch of foreign bodies and environments—played an important role in generating public optimism in the imperial state as protector of life. The iconic imagery of infectious, disabled black and brown bodies helped to mobilize hopes that state and market forces could control national vulnerabilities by managing interspecies environmental circulation. *Dread life* is the term I use to describe this racialized channeling of the fear of infectious disease into optimism regarding the remaking of life through technical intervention. This conjunction of fear, anxiety, and hope appears in the two epigraphs to this introduction from Harlan and Rumsfeld. Their statements publicize similar fears of epidemic smallpox attributed to racial geographies of risk, but they also indicate how a disease prevention technique like vaccination had been normalized as a humanitarian technology. Indeed, the implementation of childhood vaccination and other public health measures in the twentieth century significantly reduced the incidence of harmful diseases and extended the human life span across many national populations. Nonetheless, the historical conditions under which such interventions were deployed have often been guided by a combination of geopolitical and medical concerns deeply imbricated in capitalist geographies of racial differentiation and changing relationships between states, publics, and species. Understanding these links requires an account of race's materialization through a number of domains of public activity, including the proliferation of media forms, the intervention of health authorities, and the organization of public space through technology and settlement. While it is important to learn from discursive analyses of race that analyze the content of racial language and images, this book suggests that the tenacity of race also inheres in its fluid materialization across different contexts and forms of circulation, including circulation through and across human bodies inhabited by viruses, bacteria, and animals incorporated as food and medicine.

In the case of infectious disease, the public articulation of race often operates through logics of profiling, which aggregate various phenotypic and epidemiological cues into thresholds for state intervention. Although race is not "real" given that racial sciences cannot convincingly distinguish phenotypically constructed races as genetically identifiable populations, it does operate through actual material sites of bodily reproduction and contact that are subjected to surveillance in moments of crisis before they recede back into the grain of the everyday. Such ebbs and flows of racial securitization work to mask the persistence of race in the structures of governance. In the words of Arun Saldanha, "race is devious in inventing new ways of chaining bodies. Race is creative, constantly morphing . . . , but mostly it is fuzzy and operates through something" other than a sovereign structure of apartheid.[8] This book contends that racial differentiation of epidemiological risk takes particular forms that respond to contexts of military, economic, and territorial expansion as well as to specific technical logics aimed at controlling infection or burnishing national immunities.

The remainder of this introduction outlines some conjunctions of race and species in processes of imperial securitization, working through the diverse and divergent social and technical logics of disease interventions that render the body a site of vital warfare. Within such conjunctions, "security" masquerades as an apolitical good since it is experienced as the letting loose of public freedoms, allowing the disavowal of security as an agent of inequality of mobility and accumulation. As Michel Foucault contends, "freedom is nothing else but the correlative of the deployment of the apparatuses of security," experienced as "the possibility of movement, change of place, and processes of circulation of both people and things."[9] Thus the "sacrifice" taken on by the vaccinating subject in Harlan's and Rumsfeld's statements internalizes the sentimental identification of the vulnerable soldier as the potential mobility of the nation, making vaccination appear as a universal form of medical improvement rather than a technology of empire. This means further that the minoritized subject or population is often the subject of incorporation rather than exclusion; health discipline is often provoked by the specter of racial difference even as the biological work of containing risk means domesticating foreign bodies and subjects into the imagined immunities of the settler nation.

These dynamics have been intensified with the ambitious forms of intervention developed by infectious disease experts who took advantage of strong state authority to control land, markets, and institutions in US-occupied territories outside of the continental borders. Over the course of the long twentieth century, their public health projects transformed the problem of contagion from one of containing predictable risks through spatial techniques—border quarantine, environmental eradication, and the development of segregated hospitals and camps to isolate the ill—to diffusing control of risk into pharmaceutical markets and eventually into state-run vital systems that attempt rapid, large-scale delivery of medical and environmental intervention. These developments signal a shift from an association of contagion with specific geographic sites to an anticipatory governmentality that disperses control of bodily contact into the control of immune time and scale. This in turn led to more expansive forms of speculative risk assessment by the late twentieth century. Despite this general historical trend, the dispersal of disease intervention from spatial targeting to the management of time and scale did not signal the end of public health's territorial emphasis. The security state's emergency powers to control the movements and contacts of the contagious body remained in place throughout the long twentieth century, even as spatial control was layered through more diffuse and complex methods of containment couched in the optimistic promises of modern medicine.

Governing Bodily Transition

Infectious disease is normally understood as a pathological state of hybridity in which microbial species occupy and reproduce within the bodies of larger species (including humans). From this perspective, a disease-causing microorganism is a parasite that threatens the body's functions, even life itself. Yet when we take a step back from the spectacular imagery of contagion as a deathly and debilitating pathology, it is possible to glimpse the queer potential that disease reveals as immanent to life itself. To call life itself "queer" is to dislodge it from commonplace pro-life discourses that compress biological life into the able-bodied, individuated, and anthropomorphized reproductive form of the body.[10] The greatest error of anthropomorphism is not that animals or objects are incorrectly attributed human characteristics; it is that those viewed as transparently human are extracted as such from their constitution in a broader domain of biosocial life. Ed Cohen helps us understand the scale politics that anthropomorphize the human, that imagine that human bodies are self-contained and stable in contradiction to their entanglement with and constitution through other bodies. For Cohen, "The politics of viral containment relentlessly plays upon the contingency of the human 'we.' It conceptually and materially confounds our understanding both of how individuals constitute our collectives and how we exclude other collectivities . . . whether these 'others' are other individuals, other populations, other humans, other species, or other non-vital entities."[11] At minimum, life writ large exceeds the parameters of the living done by any body or population marked by species, race, or other biologized category; in this book when I refer to "life," I mean the interconnected life that crosses organisms, species, and technologies. Mel Chen, however, prefers "animacy" rather than "life," for the interaction of all sorts of matter from rocks to plants calls into question the assumption of the holism of mammalian bodies. It enlivens the apparently dead bodies of minerals, toxins, viruses, and other objects whose affective traces are mobilized in discourses of environmental risk or which prosthetically extend the body's capacities to affect others in time and space.[12] Fears of epidemic disease highlight the shared animacies of species and matter, demonstrating the transitional nature of bodily entanglements.

The body's transitional form—its plasticity that may be accelerated by disease and other forms of interspecies contact—is a significant cause of concern for securitizing states. Fear of transition is often the justification for public health interventions such as the vaccination campaigns refused by Jacobson and promoted by Rumsfeld. Yet in instituting vaccination, the state may in fact activate other forms of transition, shifting the very biological character of national immune ecologies. Under what conditions are forms of bodily transition experienced as integrated into the linear, everyday unfolding of life toward death? When do they provoke dread and the force of containment, the attempted remaking of life? And under what terms are they actually posited as the proper solutions to collective vulnerability? Drawing on critical approaches to embodiment in queer, feminist, disability, and species theories, this book asserts that the bodily changes we experience through states of disease and immunity reveal some of the transitory possibilities of vital matter. Trans studies scholars, especially those working at intersections with feminist science studies and disability studies, have noted that transitional embodiments, including but not limited to gender transitions, are extremely common across species despite pervasive social discourses that stigmatize transition.[13] The problem is that transitions including disease are policed. Living within neoliberal regimes of security partly means living in prognosis, self-managing transition between processes of living and dying by monitoring indicators of how and why a body lives longer or shorter, healthier or sicker, able-bodied or somehow disabled.[14] Increasingly, it also means living against the temporality of dread life, life affectively oriented by incalculable vulnerability that exceeds scientific means of prediction and permeates capitalist crises of futurity.

Yet fears of contagion often visually and narratively circulate through media in ways that contain risks to empire. Even as they persistently exaggerate risks of contagion, American-exceptionalist disease representations invest hope in the dominant forms of state vision and the apparently homeostatic tendencies of capitalist systems. This reflects dread life's heightened speeds and intensities of crisis representation that (1) posit the environment as an unruly site of perpetual risk, and (2) shore up an imperial optimism in the force of the state that tends to far outstrip its actual ability to control interspecies and transborder relations. Even as public depictions of contagion provoke fear that is disproportionate with actual risk, they tend to converge with reassurance of the ultimate security of settler ways of life. Priscilla Wald contends that popular US "outbreak narratives" follow a conventional structure: they identify an unknown vector of disease, trace the geographic dispersal of infection, locate a carrier, and conduct the epidemiological and laboratory labor that ends with the containment of disease through established scientific and public health institutions. Wald takes pains to document the forms of stereotyping that travel along the geographies traced by outbreak narratives. The industrial working classes and racialized migrants play key roles in the narrative production of suspicion and suspense when their supposedly risky bodies and styles of living come into contact with global capital networks that can conduct biological matter at increasing speed.[15] In the worlds of the outbreak narrative, the spiraling bodily risks of circulation and touch conventionally bring about their entropic solution, as globalization supplements new risks with endless innovation. According to such imperial representations of disease, complex systems inevitably find homeostatic fixes to the disorders of positive feedback they produce.

In similar fashion, outbreak narratives obscure the important ways in which nonhuman animals are entangled in forms of government that attempt to manage bodily transition and risk. Farmed animals and parasitic insects emerging from the peripheries of empire — for example, mosquitoes carrying malaria at the Panamá Canal Zone a century ago or Thai chickens incubating avian flu today — conventionally appear in US media as sources of zoonotic (animal-to-human) disease transmission. Animals thus rarely appear in popular culture as sites of successful immune optimization. More often than not, the figures of the vivisected, the farmed, and the wild animal all conjure the gothic risks of interspecies contact. This association of animality with primal contagion, an ongoing colonial anxiety that tinges various moments of globalization with fears of uninvited bestial intimacy, obscures the commonplace ability of pharmaceutical markets to effectively transform animal bodies into immune biovalue that is unequally distributed across wealth inequalities, health systems, and intellectual property regimes. For example, in both Harlan's and Rumsfeld's statements quoted at the beginning of this introduction, containment of infectious risk was not only a matter of exercising police or military force, but was vested in privately manufactured animal-derived drugs like smallpox vaccine.[16] Meanwhile, in contrast to these species optimized for laboratory production, states may also subtly proliferate or let loose the energies of other species such as insects that transmit zoonotic infections to precarious groups of humans outside zones of securitization. In such interspecies processes governing bodily transition, power cycles across biopolitical divisions of human/animal, body/environment, and settler/alien.

This book's narrative of what I call the *government of species*, then, does not assert the actual ability for empire or the human to fully dominate bodily transitions or immune ecologies, but rather to explain how empire takes on life as a field of potential intervention. My account of the government of species thus adds a dimension to emerging animal studies scholarship that often views humans and animals locked in a Manichean drama of species war. Such accounts are scrambled when microscopic species—species that are contagious and that occupy humans and other animals alike—demonstrate how animal and viral life can be mobilized and contained in ever more complex techniques to securitize life. The idea of the human's domination of life on Earth—featured in narratives of progressive medical modernization as well as in some liberal versions of animal rights and environmentalist politics—is an ideological obfuscation. The world history of disease control is, for the most part, the history of a failed dream of species extermination.[17] The idea of the government of species implies not just the ability of states to intervene in the immanent relations that constitute life, but also the more curious possibility that the energies, forms, forces, and adaptations of species in some sense govern the normatively anthropomorphized space of politics. As such, this book temporarily suspends normative judgments about the threats posed by viruses and bacteria. This is not only in order to overturn anthropocentric assumptions about body-environment relation, but also to understand how neoliberal immunological knowledges deployed in the service of empire today mobilize such judgments in order to justify the interventions and interests of powerful state and capital forces. Thus despite painful economic and political inequalities that make humans differentially vulnerable to environmental disasters like epidemics, viruses and bacteria must be approached as world-making species rather than simply as dangerous pests demanding eradication. In jumping scale to reproduce with and through mammalian bodies, they show how life lives otherwise, including the transitional life that we conventionally assume is contained within the space of the skin and rendered linear in the chronotope of the human lifetime.

Immune Ecologies

With the hindsight of over a century of intervening history, the post-*Jacobson* order of disease control entangles state power to intervene in the field of life with increasingly complex immunological accounts of the relation of disease to environmental factors and geopolitical conflict. While Harlan's vision of citizen sacrifice in the *Jacobson* decision reflected a world in which experts anticipated a direct relationship between space and contagion, the discovery of "Typhoid" Mary Mallon as a healthy carrier of disease in 1907 opened up new uncertainties concerning the state's ability to distinguish ill and healthy bodies, sanitary and infectious spaces. As public health officials and medical researchers investigated the ways in which disease was transmitted invisible to the human eye, ecological models of disease explicitly explored problems of settlement, urbanization, and networking across colonized space. Enmeshed in processes of territorial, economic, military, and biological expansion, disease ecology gave a specific form to the matter of the immune system that stressed its constant emergence through interspecies entanglement. While in the early phases of bacteriology at the turn of the twentieth century, the dominant military metaphors of disease conceptualized bacteria as invading the immune defenses of a stable and self-contained body, disease theorists moved on to tropes of complex ecological systems and, later, informational immune bodies in order to understand how disease transformed both body and environment in time and space. In the process, the emergent neoliberal immune system became decidedly inhuman, laboring to recognize or encode endless potential microbial threats as well as to constantly recode the self-recognizing human individual to be defended.

Networked immunity, a concept that is today central to the biological front of the US "war on terror," emerged over the course of the century following *Jacobson*. While its roots were in early twentieth-century discoveries regarding the complex activities of immune system cells, networked immunity was explicitly theorized by Danish Nobel laureate Niels Jerne and his colleagues beginning in the 1970s. They explored ecological relationships between antibodies (the primary cells of the adaptive immune system) and antigens (any particle encountered by the immune system that produces an antibody response). Jerne proposed a network theory of immunity, building on research suggesting that the receptors of lymphocytes (a key type of antibody cell) simultaneously code to attract and engulf multiple antigens including other antibodies of the immune system. Since antibodies (immune cells) can themselves be coded as antigens (potential threats) that provoke other antibodies to respond and engulf them, the immune system does not only attack externalized threats like infectious viruses or toxins, but engages in a complex play of attraction and engulfment of itself. This theory scrambles hard-and-fast body/environment distinctions. Concepts of networked immunity led immunologists to view immune response through paradigms of homeostasis and complexity, even to theorize a grammar of immunity based on language's assimilation of infinite differences into systems of meaning.[18] According to this logic, there is no internalized immune response to external environmental risks, but rather the serial production of ideotypes by which immune systems continually reprocess autoimmune recognition. Put more simply, the immune system does not so much recognize an already-formed self in opposition to outside threats. Instead, it repeatedly re-creates the borders of the body through the constant cuts it makes across the microbiome. This iterative performance of self-recognition thus inevitably involves minor processes of self-destruction. Against historical conceptions of political and bodily immunity as an exercise in the abjection of external threats, twentieth-century networked immunity becomes an exercise in ecological balance and a problem of reproducing the self by managing the complexity of its interspecies relation.

If such theories do not explicitly conjure the racial or the colonial, Warwick Anderson claims they were based on a longer history of settler-colonial disease theorizing by the likes of F. Macfarlane Burnet, René Dubos, and Frank Fenner. Jerne himself repeatedly cites the work of Burnet, who originally theorized that antigens select antibodies within the immune system, reversing the established causality positing that immune cells targeted and disposed of foreign bodies. Burnet depicted European and Euro-American soldiers involved in the conquests of the Spanish-American War as invasive species, suggesting that for them to succeed in acclimatizing white settlement in the tropics, it would be necessary for their bodies to effectively and quickly adapt to the native antigens that sought out and engaged their own immune systems. Yellow fever and other microbes were rendered as indigenous bodies that were deeply integrated into ecological balances with larger species. Thus immunities themselves carried traces of racial differentiation, wherein differently situated groups appeared more or less immunologically naive in relation to a given microbial environment. Races held a kind of embodied memory of pathogens across space and time, a memory that was tested as native antigens encountered invasive settler immune systems. Anderson sums up the relation of ecological disease theory to settler colonialism succinctly: "Resident in settler societies such as the United States and Australia, pioneer disease ecologists were especially attuned to the persisting impact of colonial development policies, to the lasting effects of agricultural change and human resettlement. . . . Disease ecology thus emerged as a legacy of settler colonial anxieties."[19]

Given this history, it is possible to track the work of settler-colonial racialization of disease into the neoliberal deployment of theories of immune complexity, which problematize the settler body as an effect and agent of circulation that requires specific logics of internal and external mediation in order to securitize its free flow. Dubos, the French-born American immunologist, situated such immune entanglements in terms of the limits of North American settler expansionism and its production of a sedentary form of life, which in turn required the adaptation to complex forms of circulation and massified forms of bodily contact. For Dubos, "North America provided an ideal setting for the euphoria of the nineteenth and twentieth centuries, which was based on the belief that industrial civilization would inevitably generate happiness by increasing comfort and by creating more, better, and cheaper goods. The vastness and emptiness of the continent made it easy for the settlers to accept the myth of the ever-expanding frontier. . . . After the whole continent had been occupied, the explosive development of science and technology provided grounds for even greater optimism." However, for Dubos, the postwar "urge for economic growth has been increasingly overshadowed by public concern over the undesirable consequences of development: crowding, environmental pollution, traffic jams, surfeit of goods, and all the other nauseating and catastrophic by-products of excessive population, production, and consumption."[20] Dubos's musings on the risks of systems and the environmental problems that scale up with neoliberal society reflect a connection between colonial understandings of immune ecology and the emergent neoliberal logics that cast the immune system as engulfed by the intractable risks of transborder contact.

Affective Warfare

These shifts in immunology help us understand why Rumsfeld stoked public fear of a virus that is today extinct in the environment. It is the very absence of smallpox from everyday life that mobilizes defense planners' fears that an errant laboratory sample or long-lost bioweapons cache will reincarnate the virus in a new environment with declining immunity. Extracted from the ecological balances that Burnet theorized as vital to successful settlement, "smallpox" meant something different to Rumsfeld than it did to Harlan a century earlier. In the taxonomy of risk Rumsfeld used to justify the Iraq War, smallpox constituted the "unknown unknown," the ultimate catastrophic risk that had to be profiled and preempted through full-spectrum warfare.[21] National infrastructures can be used to modify and burnish national immunity, constituting the state itself as a node of immune environments.

Such anticipatory containment is central to the emerging logics of biosecurity and biowarfare. The *Oxford English Dictionary* dates the word "biosecurity" to 1973, and it is often defined narrowly to refer to the capacity of laboratories to contain the potentially deadly effects of the storage of and experimentation on dangerous microbial pathogens. More broadly, the logics of biosecurity took shape in US defense circles after World War II, when defense planners developed more elaborate forms of second-order reflection upon the vulnerabilities of the state. As an outgrowth of the collision of imperial tropical medicine, public health infrastructure, and sophisticated state planning documented by Andrew Lakoff, biosecurity preparedness has catalyzed new speculative aesthetic modes for generating affective knowledge of catastrophic risks to body and nation.[22] With decreasing focus on statistical probability of risk and increasing fear of catastrophic harm, emergent forms of biosecurity raise questions about representation and the management of public fear. What media assemblages emerge to help give form to the formless specter of the outbreak?

The chapters of this book explore a history of disease representations that persistently racialize populations and environments associated with disease via the gothic figures of the alien, the rogue, the vampire, the zombie, the monster, the terrorist. These inscrutable figures hold intimate, even eroticized threats: the invisible weapon, the body poised to explode into violence, unthinkable knowledge, improper touch, unknown penetration, unpredictable shape-shifting. Yet understanding the ways in which such figures generate political crises or intensify public feelings of fear and hope require situating them within broader ecologies of representation, attending to how they circulate in media as well as how they cultivate specific forms of governmentality. Content is not enough—it is necessary to understand how these representations are circulated and how they relate to the materialization of public space and technology. Unlike most established colonial histories of medicine and public health, I argue in this book for the need to account for orders of representation that cross the subject through the affective, that shape the forms of interface available to humans, animals, and viruses, and that subtly vest governmental force in the lifeworlds of interspecies contact. To argue that the political takes the form of life is also to embed representation in life. Emerging studies of affect, invigorated by biosemiotic accounts of interspecies intimacies, can further illuminate this relation.[23]

Affect—the capacity of machines and bodies to affect and be affected, to sense, interact, connect, differentiate, move, and transition in a lifeworld—is an increasingly significant topic of debate in social theory given the regularity with which bodily sensations and public feelings are cultivated in media, in war, in technology, in formal political debate and organizing, and in the organization of social space. Affect at times appears in social theory as a taxonomizable set of capacities of the physical body. This leads to understandable criticisms of affect theory's essentialism.[24] Affect has also been historically tinged by the sorting strategies that conjoined animals, women, and the colonized outside of the sphere of the reasoning human, controlled by somatic impulse.[25] Finally, there is the problem that affect at times appears to have no specific content or defining characteristic; theories of affective labor call this labor "immaterial," while the figuration of affect through a surface/depth binary at times posits that affect consists only of fleeting surface intensities.[26] Yet these concerns do not mean that affect should be dismissed as a nostalgic return to body essentialism or a Cartesian removal of mind from body.[27] In vitalist thought (often associated with philosopher Gilles Deleuze), affect refers to the unformed intensity and potential of connection crossing bodies and living systems.[28] This contrasts with behavioralist and psychoanalytic variants (theorized by Silvan Tomkins and Sigmund Freud, respectively), where affect narrows to the nervous system and gives an almost mechanical form for bodily reaction to stimulus. Affect might be more usefully treated as a deep domain of complex bodily processes for signaling, contact, and reproduction rather than a kind of physical "kernel" underpinning emotions. While some critics may interpret this insistence on what Brian Massumi calls "the autonomy of affect" as enforcing an essentialist mind-body split, the point is not to separate mind and body but rather to give due diligence to the complexity of embodiment that colonial knowledge projects have historically denied it.[29]

To that end, I explore affect as a domain of vital potentials that not only encompasses the forms of interface situated in the nervous system, but explicitly connects the materials of the nervous system (which interfaces subjects with mass-mediated texts and images) to immune and digestive systems (where bodies are constituted through interspecies assemblage and biotechnical intervention). Alimentary, immune, and nervous interfaces connecting bodies to media cultures often shape empire's racialized sites of bodily fear and disgust. They also offer the potential of the subject's transformation through the autoimmune incorporation of alien microbes, the ingestion of animal-derived pharmaceuticals, or the generation of affective knowledge through health education or biosecurity preparedness exercises.[30] The transfer of affective potential from digestive and immune systems to mass-mediated publics (the mass national nervous system) and back may follow any number of different pathways, through the laboratory, media, environment, or the body itself. Affect has spatial dimensions, wherein it is subjected to forms of biopolitical uptake accelerated by risk media. Understanding this complexity connecting the public feelings of risk and security to various bodily systems requires particular methodological strategies. The chapters of this study often examine raced and gendered depictions of epidemic crises, detailing their uptake in particular state apparatuses or health institutions. In other moments, it is necessary to turn to phenomenologies of bacterial and viral migration to explain how racialized depictions of disease relate to particular bodily sites of transition and vulnerability that burnish public fear of contagion. Finally, the book situates these sites of affective interface within geopolitical analysis of territorial, military, and economic expansion. Such methods cross the interdisciplinary terrains of transnational cultural studies, feminist science studies, colonial medical history, political and environmental geography, and queer phenomenology.

Government of Species: Managing the Space, Time, and Scale of Epidemics

The chapters of this book build on three decades of scholarship analyzing empire's "medicalized nativism" in postcolonial studies and American studies.[31] At the same time, they move in new directions by (1) incorporating historical and geographic comparison across sites of US intervention, and (2) retheorizing empire's political form via critiques of species, disability, security, affect, and reproduction. The chapters assemble a diverse set of literary, visual-culture, scientific, and state archives of epidemics, disease control projects, and health activisms connecting Boston and Baghdad, New Orleans and Panamá City, San Juan and Kolkata, Honolulu and Guantánamo Bay, and points between. These transborder and transpecies routes of power demonstrate a strong emphasis on US empire's constitution through vital exchange between Asia and the Americas. They also document challenges to imperial medicalization and environmental warfare by patients, prisoners, soldiers, immigrants, nationalist movements, and research animals—struggles that might seem disconnected save for a comparative account of the government of species.

Space

The first two chapters focus on the spatial control of bacterial and viral contagions at the borders of US expansion in the early twentieth century, when US settlers and military forces occupied territories across the Caribbean Sea and Pacific Ocean. Exploring the battles over Hansen's disease (leprosy) segregation in annexation-era Hawai'i and military projects to contain venereal diseases at the Panamá Canal Zone during the world wars, these chapters explain how fears of racial and sexual contact invigorated a missionary zeal for segregation and other spatially targeted health powers, even as social movements contested them. This imperative to discipline the spatial relations of built and natural environments was already affirmed in *Jacobson*, the rhetoric of which was largely concerned with the retrospective justification of state quarantine and sanitary power.

In this context, efforts to control the spread of bacteria and viruses across groups were eventually strengthened into an ideal of fully eradicating environmental pathogens, of exterminating specific microbial species in settler spaces.[32] Such efforts were intensified by fears of two forms of transborder movement. First, a series of epidemics spread in port cities and were associated with Asian and European immigrants. Benjamin Harrison ordered the exclusion and deportation of foreign-born people with Hansen's disease (then called leprosy) as early as 1888. Following typhus and cholera epidemics in New York, Congress passed the National Quarantine Act in 1893 for port screening and upheld the authority of states to indefinitely isolate those who were potentially ill or exposed. The 1902 case *Compagnie de Francaise de Navigation a Vapeur v. Louisiana State Board of Health* allowed health authorities to block the entrance of immigrants if either immigrants or the receiving community were suspected of harboring disease. A year later, in the Immigration Act of 1903, the United States explicitly barred entry to persons "afflicted with a loathsome or with a dangerous contagious disease."[33] Second, territorial expansion spurred fears of disease during the decade 1893–1903, when Guam, the Philippines, Cuba, Puerto Rico, Panamá, and Hawai'i were all occupied by US military forces. Fearing the purported failures of disease control in British India, military and philanthropic organizations established expansive health authority to control territory, enforce treatment, and conduct research and eradication, often in ways that unequally benefited settlers against indigenous populations.

Yet the ability to simply contain microbial species through segregation of human populations emerged in tension with the American-exceptionalist myths of a benevolent empire. Chapter 1 analyzes the frenetic fears and hopes apparent in the public circulation of images of Hansen's disease patients during US debates over Hawaiian annexation. Even as literary and medical images of the disabled, dark-skinned "leper"[34] were virally circulated to justify the program of segregation on the island of Moloka'i, I recount how enforced treatment complicated settler claims to institute health authority as a universal public good. The resulting battles between patient activists and the annexation-era health apparatus, which played out in journalism and literature concerning the public health programs, transformed the available treatments and medical language while setting the stage for open conflicts between Hawaiians and the settler state. This account complicates dominant necropolitical theories of the relationship between US empire and the spatial technology of the camp, which was reformed in the face of protest and sentimental accounts of suffering. Building on this analysis, chapter 2 reconstructs the history of the incarceration of suspected sex workers at the borders of the Panamá Canal Zone during the world wars. In contrast to the expansive and invasive goals of tropical medicine in Hawai'i, venereal diseases (primarily syphilis and gonorrhea) occasioned a medicalized state of war that attacked environmental space using the model of antimalarial campaigns aimed at controlling mosquitoes. This resulted in a series of new measures to regulate public sexuality for deployed soldiers. Restructuring the transborder control of human-microbial hybridity through vice policing, sanitary corridors, prophylaxis stations, and the registration or suppression of brothels, the US military and Panamanian authorities aimed at producing a mobile sanitary cordon around soldiers. This defense underwrote an offensive strategy allowing diseases of all kinds to freely proliferate outside of the Canal Zone, where Panamanians generally and women in particular were figured as traitorous, swarming embodiments of gonorrhea and syphilis. Yet because of the threat to the controlled consumption of urban capitalism, business elites quickly brought an end to the program, even as they allowed the police to violently suppress women's protests. Exploring the anxieties evident in the divided administration of such protests, I explain the constitution of Panamanian cities as "fallen" spaces of American vice and contagion, a construction evident in a variety of literary works critiquing the transformations of Panamanian life under military occupation.

Time

While the first two chapters explore disease intervention as a technology of territorial expansion in the first half of the twentieth century, by the end of

World War II an era of emergent decolonization transformed the logics of empire's territorial project and spurred a combination of military, political, and market forms of expansion that included the United States' development of international pharmaceutical markets.[35] By the onset of the thirty-year crisis of the Great Depression and the world wars, the *Jacobson* precedent would apply to a world in which the nature of disease and the outlines of empire were decidedly murky. With the knowledge of asymptomatic carriers like Mary Mallon, diseases were viewed as moving without physical signs. In response, an emergent biomedical research infrastructure invested hope for prophylactic forms of bodily and environmental discipline, especially through the circulation of animal-derived drugs through pharmaceutical markets. In contrast to the vigilant spatial control of classical public health, pharmaceuticals like vaccines and antibiotics would mitigate the risk of disease in advance of contact or at the beginning of infection rather than by targeting specific spaces for eradication. With the establishment of the National Institutes of Health (NIH) in 1930 and the Centers for Disease Control (CDC) in 1942, state infrastructures would marshal expertise and expensive research materials for drug development. Accelerated by the eventual Cold War competition in technology with the Soviet Union, there was by the late 1950s a growing state investment in national research infrastructures targeting a variety of diseases and disabilities. This backdrop to the so-called antibiotic revolution of midcentury required two displacements of risk: the recruitment of ongoing populations of human subjects for clinical trials (often under compromised conditions of consent) and the transnational incorporation of animal capital in the form of vivisected bodies and body parts.[36] The bodies of research animals were rendered as both living media for the production of pharmaceuticals and as almost-but-not-quite-human models for testing drug safety and efficacy; as such, they became "strategic material" that one CDC committee declared nearly as important to national defense "as tungsten and tin and natural rubber."[37]

Chapter 3 registers these shifts by tracking the public fear of polio and its impact on emerging neoliberal geographies of the biomedical primate trade. In this example, time itself became a privileged arena of intervention. Diseases needed to be prevented from transforming the body by burnishing immunity before contact. This was especially true for a "white disease" like polio, which was understood as having already penetrated the immune ecology of the nation. The research and production of the famous polio vaccines involved a massive state apparatus to displace direct spatial control onto indirect geographies of biomedical research resource extraction, including through the harvesting of research primates from across the colonial world in order to model diseases and develop new drugs. Notably, this meant that technical intervention reversed the xenophobic assumption that foreign materials would infect the settler nation from the outside; the colonial extraction of animal capital could in fact itself become the source of a new medical modernity, burnishing settler immunity. One major development of this shift is that quarantine and other forms of sovereign spatial intervention were displaced onto human and animal bodies required for clinical trials and pharmaceutical production. I trace these biopolitical logics from the founding in 1939 of a primate research colony at Cayo Santiago, Puerto Rico, through a variety of primate labs that the NIH attempted to establish domestically and in central Africa from 1957 to 1961. These programs, along with importation schemes to bring rhesus macaques to the United States from India for the production of polio vaccine, relied on long-standing forms of colonial enclosure and resource extraction. They also exhibited a new Cold War urgency that attempted to re-engineer the form of domestic life and to secure the material bases of immunity from the decolonizing world. This controlled appropriation of primate bodies intensified new fears of biomedical intervention, as images of transplantation and primate invasion in Hollywood film and other arenas of visual culture expressed deep ambivalence over US pharmaceutical imperialism's entanglement of state and society in the Cold War politics of decolonization and containment.

Scale

As they attempted to win alliances with third world states during the Cold War, the United States and the Soviet Union funded an "international health" movement attempting to scale up disease control across the globe. Following on the apparent success of polio vaccine in most wealthy countries, the dream of total eradication of a disease was briefly realized in the World Health Organization (WHO) smallpox campaign, which in 1977 succeeded in fully exterminating the variola virus from the environment. Public health officials and intergovernmental organizations alike began to declare that the developed world was transferring to the developing world the sanitary conditions and public health knowledge to prevent the deadly diseases of the past. Over two centuries, the average life expectancy of a human born on planet Earth had doubled to over sixty years, with that in rich nations approaching eighty years. Yet by the 1990s, WHO once again declared a growing threat from infectious disease.[38] The transnationalization of labor, increased urban-rural contact, wealth inequality, and the migration of environmental pollution meant new outbreaks of established diseases, while drug-resistant diseases emerged from the cycling of bacteria and viruses between geographic areas in which pharmaceuticals are overutilized and underutilized. In the 1980s, new zoonotic diseases such as HIV dashed the imperial fantasies of both eradication and pharmaceutical magic bullets. At the same time, knowledge of the lengthy histories of US and Soviet bioweapons development fanned paranoia among US defense planners concerning bioweapons proliferation following the fall of the Soviet Union. In this context, the specter of force outlined in *Jacobson* returned: from South Africa to the United States to China, carceral quarantine and emergency law would again become common discussion topics of public health officials. These developments coincided with the rise in the 1990s of the "emerging diseases" paradigm, which integrates ecological, economic, social, and biological factors for understanding how risks migrate in an era of neoliberal globalization. Building on ecological models of immunity, the emerging diseases worldview figures the blowback of colonial inequalities as an emerging threat to settler space, and suggests the expansion of neoliberal pharmaceutical markets combined with an incipient militarization of public health as an integrated, full-spectrum defense.[39]

The final two chapters focus on this dynamic in which public health preparedness requires scaling up capacities to interdict future risk and militarized strategies to preempt the very emergence of these threats. Chapter 4 explores the Smallpox Vaccination Program during the Iraq War and documents the roles of journalists, fiction writers, and government officials in the reanimation of smallpox as an incitement to indefinite war. As the world's only eradicated infectious disease, smallpox literally exists in the suspended animation of laboratory culture; its absent presence and reanimation has been repeatedly staged in journalism, popular novels, and defense policy from the 1990s through the lead-up to the 2003 Iraq invasion. Represented as the world's most deadly and disfiguring virus, smallpox suggests the extreme potentials of disease emergence at the borders of empire, where unruly ecologies mingle with the monstrous visions of the rogue and the terrorist. This chapter explores how visions of emergence circulate racialized fears of smallpox in the biosecurity apparatus and across to broader publics, intensifying virally through association with figures of Asian bodily excess, the specter of the American Indian genocides, and the infected bodies of rhesus monkeys confined in secret government laboratories. In the process, security institutions as well as their critics have contributed to forms of crisis governance based on expansive new

forms of risk speculation. The chapter thus provides a queer, interspecies critique of the current neoconservative doctrine on biological weapons, demonstrating how the government of species operates through the production of
racial uncertainty and the management of imperial information economies at
a moment of intensifying state secrecy.

Chapter 5 argues that behind the current turn to emerging diseases and
high-tech biosecurity lies the blunt hand of the racial security state whose
powers for health intervention were enumerated in *Jacobson*. It analyzes the
management of the space, time, and scale of disease control in the first US
national response to a so-called emerging disease: AIDS. Preparedness in this
case involved a militarized return to the camp through the emergency incarceration of HIV-positive Haitian refugees by the US Coast Guard at Guantánamo Bay, Cuba, beginning in 1991. I argue that indefinite health detention
during the AIDS crisis involved a war over the form of the Haitian refugee
body, which was subjected to a host of animal and microbial threats as HIV-
positive Haitians fleeing political persecution were resettled in a US-run concentration camp. In response, the refugees initiated a hunger strike that would
accelerate the transitions wrought by the virus, forcing the state to define the
acceptable terms of a livable life. The resulting court decision required the
parole of Haitian asylum seekers not based on any standard political or legal
threshold, but rather through medical tests that determined the strength of
their individuated immune systems. Even though the legal challenge *Haitian
Centers Council, Inc. v. Sale* eventually brought the end of the violent and illegal incarceration of Haitians at Guantánamo, the Haitian case demonstrates
how imperialist discourses of security are articulated through uneven formations of sovereignty that produce visions of health and humane care through
the targeted dissemination of precarity and death. This biopolitical formation
guides the logics of more recent state claims of "humane treatment" of post-
911 prisoners at Guantánamo, whose advertised rights to recreation, freedom
of religion, and health care are rendered ambivalent through the criminalization of death.

The epilogue explores how the control of time and space, the scaling up
of biological defense, and the militarization of interventions into animal and
viral life reframe an understanding of the tenacity of US empire in the twenty-
first century despite clear and growing limits on US power. Arguing that the
government of species illuminates a history of the present configurations of
state and social welfare in the current phase of neoliberalism, it notes the
contradictions between two key trends: on the one hand, there is an infinitely

expansionary regime of global biosecurity that attempts to incorporate an increasing number of geopolitical entities, species, and technologies into its arsenal of preparedness; on the other, austerity pressures on the state, technical limitations, patient activisms, and the unpredictable mutations and transits of human and animal bodies throw into crisis the expansion of empire's technocratic power over life. This has particular import for studies of US empire, which I argue has its own fully theorized logics of immanent intervention in spite of dominant narratives of imperial decline and various psychological misreadings of US imperial duplicity or contradiction. While such limits to power pose openings for redirecting the government of species away from its imperial designs, they also signal that such a politics cannot simply proceed from formulas that technologize populations or species uniformly, ignoring attention to today's increasingly circuitous linkages of the powers over life and death.

Decolonizing Energies

As Henning Jacobson's refusal of vaccine demonstrated over a century ago, disease interventions are commonly constrained by a variety of public and private responses, including organized patient advocacy for the reform of treatment; the simple refusal of medicine; political speech challenging expert opinion or state power; and efforts to escape institutions of carceral health control. This book tells the stories of a number of challenges to emergency health authority, including those of individuals who resisted Hawai'i's Hansen's disease segregation policy; women who escaped American vice raids in the Panamá Canal Zone; Haitian refugees who carried out a hunger strike and refused medicine to challenge their imprisonment at Guantánamo Bay; writers who argued against carceral quarantine and preemptive warfare; filmmakers who warned against the dangers of biomedical experimentation; and laboratory animals who escaped confinement in research institutions. Each chapter explores how the government of species opens unexpected sites of public conflict and affective politics. My discussions of these emergent energies raise the question of whether interspecies politics reveal tactics, however constrained, for the decolonization of life and death. Drawing on Maile Arvin's work on indigenous resistance to and through the precepts of colonial racial science in Hawai'i, I suggest that it is possible to glimpse a decolonial form of living not only in the refusal of medical authority, but also when bodies and collectives proliferate the energies of the government of species in unexpected directions.

Rather than actually reversing colonialism's processes of incorporation—the ostensible goal of decolonization—these acts regenerate vital energies and forms of medical citizenship in ways that redirect the force of empire in life.[40]

Two key anticolonial theorists, Frantz Fanon and M. K. Gandhi, have outlined paradigms of anticolonial opposition that also explore how affective politics take form at the site of the body. A Martinican-born psychoanalyst who practiced in Algeria and became that country's foremost anticolonial theorist, Fanon gives an account of the affective violence of colonial medicine: "The colonized person who goes to see the doctor is always diffident. . . . The doctor rather quickly gave up the hope of obtaining information from the colonized patient and fell back on the clinical examination, thinking that the body would be more eloquent. But the body proved equally rigid. The muscles were contracted. There was no relaxing. Here was the entire man, here was the colonized, facing both a technician and a colonizer."[41] Fanon relates this feeling of stress to both the material structuring of colonial space and the racial association of the colonized with the affectable, animalized body. Noting the "roar" of decolonization, Fanon claims, "a hostile, ungovernable, and fundamentally rebellious Nature is in fact synonymous with the colonies and the bush, the mosquitoes, the natives, and disease. Colonization has succeeded once this untamed nature has been brought under control."[42] The outcome of the nervous friction between colonizer and colonized is the decolonial rejection of settlement, building up to and including open physical resistance aimed at the institutions of the colonizer. Whereas readers of Fanon have erroneously narrowed this phenomenological description of colonial stress to an ethical defense of violence, Fanon himself depicts the affective transfer of the settler's original spatial violence into emergent forms of decolonial force. Colonial stress is then relieved in orgasmic terms: "In the colonial world, the colonized's affectivity is kept on edge like a running sore flinching from a caustic agent. . . . This overexcited affectivity . . . takes on an erotic delight in the muscular deflation of the crisis."[43]

While there are moments when the colonial frictions produce such open resistance, the chapters of this book document many more subtle redirections of power, ones that need not require decolonial theory to resort to an androcentric, eroticized, or binary model of resistance. These include the development of solidarities, images, and activisms contesting the governmentality of disease control as well as forms of circumvention that allow for bodies to navigate their direct forms of containment. M. K. Gandhi's experiments in body politics open into the complexity of such challenges, as he sought to cultivate fellowship across social and even species divides that Fanon saw to be intractable. The diasporic poetics of nonviolent striking and fasting Gandhi brought from South Africa to India sought to deploy public feeling precisely to serve political causes, though repeated failures and adaptations in method demonstrated the difficulties of effectively turning the living body into a site of decolonial tactics.[44] Hunger striking rendered the ascetic body a site of sentimental identification and masculine moral authority. Even if Gandhi never interpreted this in itself as a type of violence, his discussion of ethical duties to protect animals acknowledges, "Not one man really practices such a religion [of nonviolence] because we do destroy life."[45] Caught between attempts to minimize the interspecies violence in the everyday reproduction of the self through digestion, on the one hand, and the problems of the technocratic administration of emergent anticolonial socialism, on the other, Gandhi attempts to cultivate fellow feeling sparked by spectacles of suffering.

Gandhi's model of decolonial affect, unlike Fanon's, turns squarely to a humanist cosmopolitanism, and thus shares with the late writings of Foucault an ethical vision of "counter-conduct," a form of action that indicates a dispute of the civic order from within its own governmentality.[46] Of course, this book suggests a complex variety of responses to medical and sanitary authority. There is a wide divergence between the example of a rhesus macaque escaping laboratory space and an organized patient rebellion against incarceration, and I do not mean to suggest any kind of equivalence between the two acts when I situate them within a broad spectrum of energies unleashed through the government of species. Nonetheless, both events bring about the redirection of state powers that attempt to adapt imperial expansion to more smoothly unfold into life. I argue in several instances throughout the book that the form of carceral and market interventions against disease are adapted by the powers that be in response to the affective and economic limits posed by the human, animal, and microbial objects of intervention. The government of species thus requires consistent modification and redirection, a phenomenon that renders life itself an ongoing site of social and political contestation as disease interventions layer ever more complex technologies of containment and circulation. The energies of bodies and populations often spill beyond structures of containment, opening up unexpected effects elsewhere in biopolitical formations, demonstrating that material affects of intervention work to both incorporate and regenerate imperial form, often moving outside the confines of the direct goals or ideologies of intervention. This is the sense in which we can begin to speak of regenerative forms of living within empire, where life's queer potentials of contact reorganize force guiding biopolitical formations. Toward understanding these emerging political affects, the remainder of this book turns to specific historical contexts of US disease control projects, analyzing how the government of species emerges out of complex relations between institutions, technologies, media, and the many species that populate the worlds of imperial expansion.

CHAPTER 1

"An Atmosphere of Leprosy"

Hansen's Disease, the Dependent Body, and the

Transoceanic Politics of Hawaiian Annexation

The Hawaiians are a very affable, agreeable, and lovable people just as much so as any other on earth; but in contact with disease, all their desirable traits are seriously discounted by their lack of care, because they endanger all of us "by failing to obey the most simple rules of health, necessary for their own salvation and self-preservation." It is a most pitiable condition, evident to the most unobserving, "that an atmosphere of leprosy clings to and surrounds the unfortunate Hawaiian." Why? Because he fails to realize the danger that menaces him, apart even from the extreme receptivity of his system to the bacillus of leprosy, a condition lacking in other races domiciled in Hawai'i nei: this being an indisputable fact, then he (the Hawaiian) is the weak link in our chain of national health defense.

—ALFRED A. MOURITZ, *The Path of the Destroyer: A History of Leprosy in the Hawaiian Islands*, 1916

Physician Alfred A. Mouritz, who had served for four years as the resident doctor at the Kalawao Hansen's disease settlement on the Hawaiian island of Moloka'i beginning in 1884, published a history of the disease's impact on Hawai'i in 1916. Although the first page of Mouritz's book glosses English translations of the Bible, which in 2 Kings suggests that a skin disease "clings" to the general Naaman and his descendants, Mouritz declares, "Leprosy as a hereditary condition is of doubtful proof." *Mycobacterium leprae*, at the time called *Bacillus leprae*, was discovered as the cause of what we today call Hansen's disease in 1873. Still, well into the twentieth century, its mode of transmission, its divergence in severity, and individuals' seemingly random immunity to it made the persistence of leprosy a mystery. How, then, could Mouritz declare that "an atmosphere of leprosy clings to and surrounds the unfortunate Hawaiian"? The answer is a "receptivity" to bacteria that was both physical and moral, an irrational lack of self-care compounded by biological vulnerability. This explanation of how an infectious disease defines a racial "atmosphere"—an affective intensity that defines a group's relation to others in space—is compounded by the main epidemiological finding of the study. The book describes a series of increasingly macabre and unethical experiments Mouritz conducted on patients and at least one *kōkua* (helper or care worker) at the Molokaʻi settlement: testing the ability of flies and mosquitoes to contract Hansen's from his patients; holding communal meals and taking samples of patients' chewed or touched food for the presence of bacteria; conducting gynecological exams for sexual transmission; attempting to infect healthy residents or reinfect ill ones; even exhuming the grave of a former patient to extract bacilli from the corpse. Mouritz comes to a conclusion that ends with the title of his study: "The main entry of the Bacillus Leprae into the system of Man is through the *mouth* and *digestive tract*, and this is '*the path of the destroyer*.'"[1]

The oral opening to the alimentary canal presented an abyss of uncertain connection. Writing at a moment when increasing public knowledge of germ theories of disease advertised the body's physical interface with microscopic bacterial species, Mouritz draws the risk of disease transmission and immunological vulnerability into a racial typology that dominated US medical, journalistic, and literary representations of indigenous Hawaiians. Associating Hawaiians with unsanitary transfer of feces, saliva, and dirt carrying unwanted bacterial species, Mouritz's experimental conclusion regarding an oral pathway for Hansen's disease confirms Parama Roy's assessment that "colonial politics often spoke in an indisputably visceral tongue: its experiments, engagements, and traumas were experienced in the mouth, belly, olfactory organs, and nerve endings, so that the stomach served as a kind of somatic political unconscious."[2] As Hansen's disease literally damages the nerve endings in its advanced stages, Roy's phrasing is particularly apt for capturing Mouritz's triangulation of immune, digestive, and nervous entanglements with microbes. Mouritz trains the settler's fear of disease upon a Hawaiian subject who lacks the proper compass for navigating the environment, the nervous disgust that modern hygiene attaches to putrid smells, the exchange of bodily fluids and wastes, and other transfer points of interspecies contact. In a sense, this absence of affect indicated that even prior to infection, the Hawaiian was always already "anaesthetic," a diagnostic term used in Mouritz's day to describe cases of Hansen's that affected the nerves' proper capacity for feeling.

Like many other tropical medical experts and colonial travelers writing on Hansen's disease at the turn of the twentieth century, Mouritz combined modern tropes of filth, contagion, and animality with a tragic sense of the biblical antiquity of the disease. In asserting the spiritual taint that clings to the colonized and the curse of their inevitable physical isolation once afflicted, such writings actually contributed to a new and intense practice of spatial quarantine of Hansen's disease in an era of colonial tropical medicine. Against the presumption that the isolation of Hansen's disease patients was an unchanging historical practice of abjection, in the late nineteenth century the disease became an intense preoccupation of British and US tropical-medical experts crossing South Africa, India, the Pacific Islands, California, and Louisiana, despite the fact that it is only moderately contagious and slow to progress in the body.

This chapter suggests that sensational representations of native Hawaiian (Kānaka Maoli) susceptibility to the Hansen's disease bacterium, such as Mouritz's, channeled racial fears of colonial movements of labor, commodities, and settlement into an emergent government of species reliant on segregation of the ill. The technique of racial engulfment through quarantine in turn advertised the inevitable biological and political dependency of Hawaiians on the settler institutions of the US occupation. Documenting how such public fears and hopes were conjoined in literary, photographic, and epidemiological depictions of the Hansen's disease settlements on the island of Molokaʻi, I suggest that this logic of dread life invoked persistent contrasts between living and dead, primitive and modern, human and animal, and dependent and independent bodies. The interspecies transitions of disease—and Hansen's disease was, again and again, represented as a terror-inducing transition state—were narratively and visually organized through the unveiling of the purported animalization of individuals' skin, faces, and extremities. Within the resulting imperialist discourse on "the Hawaiian leper," Hansen's patients were not simply excluded from the law, but engulfed by state practices of quarantine and medical uplift in which they were progressively distanced from an ideal of American liberal individualism. If this meant the increasing use of the state's health policing powers, it also occasioned public battles over treatment and connected Hansen's patients in struggles crossing expanding borders of colonial control.

Observing the resulting patient activisms for expanded forms of medical and legal citizenship, I conclude that the case of Hansen's disease quarantine requires a reevaluation of necropolitical theories suggesting that the medical camp and the practice of human vivisection become privileged sites for the articulation of the state's power to kill. Instead, the spatial form of Hansen's disease quarantine reveals the potential for liberal reform to capture and mobilize life, to govern both the biological entanglements of species and the affective entanglements of state and subject. The stories of activists including Kalauikoʻolau and Olivia Breitha who challenged the Hansen's disease segregation policy demonstrate these conjunctions of politics of life and death as Hansen's disease incorporated Hawaiians and settlers into the architectures of public health in its golden era at the turn of the twentieth century.

Racializing Risk: Asia-Pacific Geographies of Hansen's Disease

At the turn of the twentieth century, one word dominated discussions of Hawaiian annexation in the journalism of the continental United States: leprosy. A wide variety of press reports, literary texts, and medical and governmental debates concerning Hawaiʻi in the annexation years constructed the bacterial infection as a central problem in the relationship between the islands and mainland. While annual cases of the disease in Hawaiʻi were at any given time several hundred higher in number than the total for the continental United States during this period, Hansen's disease was by no means geographically confined to one territory of Euro-American settlement. In fact, in Louisiana and the north central states of the United States, the disease had long been endemic (at times associated with Scandinavian immigrants), yet occasioned little popular representation or major public efforts at containment until the debates over Hawaiʻi produced a national policy of isolation. What accounted for this difference?

Beginning in the 1870s, industrial expansion and its unequal concentration of wealth in corporate monopolies allowed US capitalists to emerge for the first time as a financial rival to Britain. Seeking expanded investment westward in the Pacific at the moment of the closing of the frontier, the United States funneled hoarded capital into territorial expansion led by filibustering capitalists.[3] This expansion involved taking over land in the Pacific and the Caribbean, with the broader goal of the enforced opening of Chinese and Japanese markets to US trade. While fears of imperial overstretch and the dangers of expanding contact permeated public discourses on expanding territorial settlement and trade in both the Pacific and the Caribbean during the 1890s, Hansen's disease offered a ready trope of empire's racial contagion that could be transmuted into attempts to control personal and environmental space in an era of shifting borders. The risky touch of the so-called leper suggested the endangerment of able-bodied whiteness in the face of expanding networks of transoceanic trade and sovereignty. The body of the Hansen's patient was represented using the figures of beasts and zombies haunting the settler-colonial project.[4]

The first known cases of Hansen's disease appeared in Hawai'i in the 1830s, presumably brought by increasing US and European shipping activity. Despite the fact that Hansen's disease impacted fewer individuals than other diseases such as tuberculosis and spread relatively slowly, it was publicly attributed to the supposed social and sexual impurity of the Hawaiians, and thus could be disavowed as an agent of the colonial genocide perpetrated against indigenous peoples of the Pacific. Under an 1865 law of the Hawaiian Republic, signed by King Lot Kamehameha under pressure from US and European advisors at the Hawaiian Board of Health, persons diagnosed with leprosy were denied rights to property, movement, marriage, and legal standing. They were segregated in state hospitals and, eventually, in the Kalawao and Kalaupapa Hansen's disease settlements on the island of Moloka'i. Although some of their rights were later restored following public protest, institutionalized quarantine rapidly medicalized the disease and brought heightened surveillance. In the first three decades of the policy, those quarantined were almost exclusively native Hawaiians. In addition to subjecting Hawaiian "leprosy suspects" to enforced police screenings for skin lesions and other symptoms, those infected were criminalized and termed alternately and ambiguously in early Board of Health documents as "prisoners" and "patients." The disease, commonly known in Hawaiian as *ma'i Pake* ("the Chinese sickness," associated with indentured plantation laborers), would now also be known as *ma'i hookawale*, or "the separating sickness."[5] Those diagnosed with leprosy were sent to a small section of the island of Moloka'i bounded on one side by the sea and on three sides by steep cliffs. Prior to the establishment of Kalawao, the first settlement, this site was inhabited by small numbers of farmers from the adjacent valleys.

As segregation numbers grew to 3 percent of native Hawaiians (some 8,000 Hawaiians were eventually interned), pressure from the public, religious organizations, and settlement residents ensured that the government and religious groups would administer medical care, provide supplies and shelter, closely monitor patients, and develop social, cultural, and police institutions. After charges of harsh arrest tactics and wrongful detention, the Hawaiian Board of Health eventually introduced procedural rights through which defendants could challenge a leprosy diagnosis — but not the actual authority to detain. In contrast to the image of the helpless leper that continues to influence historiography of the settlements, much of the initiative behind organization and improvements in care was taken on by residents themselves, who organized a church, established informal administration, spoke out privately and publicly against the abuses of some superintendents and other administrators (including the famed Catholic missionary Father Damien), and even protested US annexation of Hawai'i.[6] Patients remained, however, subject to police power to possess property, restrict movement, and chemically sanitize both their bodies and any items they attempted to send out of the settlements through the mail. Although the segregation order in theory applied to any individual diagnosed with the disease, reports indicate that whites benefited from racial profiling, racially unequal enforcement, and the settler government's policy of allowing nonnatives to return to their home countries for treatment.[7] The vast majority of those interned at Moloka'i were Kānaka Maoli. One individual who was arrested along with his daughter for transport to Moloka'i described the segregation practice as a blatant form of state racism: "These actions are like Sweeping away the Brownskinned people."[8]

Following the 1893 US-backed corporate coup against the Hawaiian monarchy and the official 1898 US annexation, Moloka'i settlement residents did not obtain significantly expanded rights; despite the institutionalization of social services by missionaries, public health boards maintained strong policing powers, and annexation debates strengthened the regime of carceral quarantine. In 1904, the US District Court in Hawai'i declined a legal petition for habeas corpus, which would have restored the legal rights of interned Hansen's disease patient Mikala Kaipu. This was the main legal challenge to the application of the *Jacobson v. Massachusetts* precedent (which upheld enforced quarantine and vaccination) to patients segregated on Moloka'i. During the appeal to the Supreme Court, Kaipu died and the case was vacated.[9] The court never explicitly discussed the detention of people diagnosed with Hansen's disease, but subsequent legal cases such as that of Typhoid Mary in New York confirmed state authority to detain individuals on public health grounds. Meanwhile, the United States opened a Hansen's disease hospital in the Philippines immediately after the 1898 takeover there.[10] And as federal law restricted "transportation of lepers in interstate traffic" and health boards in San Francisco, New York, and Louisiana mandated segregation, US health officials began floating the idea of making Moloka'i the national leprosarium.[11]

It is no surprise then that many Hawaiians were skeptical of representations of themselves as diseased or contagious.[12] Pennie Moblo details the various ways in which Molokaʻi residents protested their stereotyped association with disease and sought forms of patient control.[13] Like other indigenous Pacific peoples, Hawaiians were subjected to both settler occupation and exposure to epidemics as twin elements of a genocidal strategy to colonize land beginning in the early decades of the nineteenth century. Yet this awareness of disease as an effect of settlement and trade forces was not acknowledged by the US press, which highlighted the seemingly inevitable demise of the native Hawaiian due to incomplete development and biological and cultural weakness. The Hawaiian was often rendered as a sentimental and biologically vulnerable victim of a disease brought by other racialized groups whose excessive bodies endangered the islander. Susceptibility of the native Hawaiian to leprosy was often contrasted to the contagion of migrant Chinese entering the Hawaiian plantation; in the words of author Charles Warren Stoddard, "It is . . . understood that the seed of the dreadful malady came from Asia, and came in the person of an ill-fated foreigner."[14] Hansen's disease was often associated with Chinese Americans in the late nineteenth century and used to justify port quarantines as well as the exclusion, segregation, and repression of Chinese immigrants.[15]

Before US annexation, a number of racial and ethnic groups were associated with Hansen's: Chinese, Japanese, and Indians or "Hindoos" were often named most susceptible, with Polynesians and Africans following close behind. Theologian Albert Palmer speculated that Polynesians were ancient descendants of South Asians in order to explain their susceptibility to disease. Including Pacific Islander, South Asian, Middle Easterner, and Chinese within a single category of epidemiological risk and a single sphere of incomplete development, he saw the US public health role in Hawaiʻi as a way of spreading Christian good faith to all of Asia, and especially the Pacific Rim markets of Japan and China.[16] Stoddard's travel narrative *The Lepers of Molokai* included a historical survey of the disease tracing its movement across the colonized world from North Africa and the Middle East to Hawaiʻi. India entered into US debates over Hansen's disease as an example of the ways in which a failing colonial administration—that of the British Raj—could allow a disease like Hansen's to spread unabated due to poor institutionalization of health and sanitary measures. (Often this was attributed to unsanitary vaccination procedures, an epidemiological theory that Mouritz disputed.) As regions with large populations and British colonial labor diasporas, China and India were posited as the originating source of and human agency behind the spread of

Hansen's disease. Hawaiians were transformed into contagious transmission points whose vulnerability to bacterial and viral hybridization opened the potential of epidemics to the settler nation. The US media addressed Hawai'i's incorporation through the lens of its potential threats to bodies on the mainland; Hansen's disease was "the most volatile issue" in the debate. The *American Monthly Review* ran an article titled "Shall We Annex Leprosy?" In an article in the *North American Review*, a New York doctor predicted that Hawaiians with Hansen's disease would enter the mainland in droves to escape quarantine at Moloka'i.[17] The leprous Asia-Pacific body was seen as excessive and potentially uncontainable.

In 1902, Surgeon General Walter Wyman presented "A Report Relating to the Origin and Prevalence of Leprosy in the United States," commissioned by the 1899 Congress to survey the prevalence of Hansen's in the United States and to evaluate an earlier proposal to make the Hansen's settlements at Moloka'i into a national leprosarium for all US territories. This proposal was later defeated by public resistance in Hawai'i as well as concern for Anglo-American patients who would be forced into faraway exile. But with expert enthusiasm for the proposal in 1902, the report's final page, written by the chief US Marine Hospital Service quarantine officer of the Hawaiian Islands, claims, "The number of lepers the place is capable of accommodating is practically without limit, and it occurred to me more than once that a site so suitable and isolated should be made more use of—that is, made our national leper sanitarium."[18] The proposed national Hansen's hospital had to make people with Hansen's disease "patients" (not prisoners) who "must not be made to feel that they are under any restraint" (10). Quarantine operates in the service of the common good, for it protects those at risk of contracting the disease and restores the dignity and humanity of the infected themselves, who, according to contributor R. D. Murray of the Key West Marine Hospital Service, "shun people instinctively" for fear they will transmit the disease (94).

The report mentions a wide spread of nationalities that highlights the presence of the disease among US Americans, Europeans, West Indians, Chinese, Japanese, and "Kanakas" (Kānaka Maoli, native Hawaiians).[19] Anticipating the logic of the *Jacobson* decision, the report presumes that the majority of cases among US Americans have been contracted outside of the country. Any discussion by a patient of travel beyond the continental borders is taken as evidence of the importation of the bacterium; even in cases with no history of travel, the report ignores the endemic status of the disease in Minnesota and coastal Louisiana to assert transborder transmission:

While your Commission does not wish to discredit the accuracy of the information furnished by the observers who have reported this large proportion of the cases as having contracted the disease in our country, it feels justified in expressing the opinion that some of them, perhaps, brought the disease with them from foreign lands. . . . [When cases supposed to originate in the United States] are examined more closely the fact is often brought to light that they spent a portion of their time in China, Hawaiian Islands, West Indies, or other places, where the disease prevails in epidemic form. (8)

The Marine Hospital Service officer stationed at Hawai'i in 1893 to report on Hawaiian Hansen's sees its origin in contact with other Pacific Islanders, Chinese, or the "mixed crews—negroes, black and white Portuguese, and Chinese—of the whalers" (96). The report calls for the segregation of immigrants, but sees the ultimate export of the US quarantine system to Asia as the only solution to the supposed crisis. H. M. Bracken, the secretary of the Minnesota Board of Public Health, explains, "The feeling that we can quarantine against lepers by watching immigrants is an unsafe one. The family history of all immigrants from a country where leprosy prevails should be secured before they are allowed to embark for America, and no member of a leprous family should be permitted to land upon our shores" (47). Bracken writes further, "we have been constantly importing leprosy," and thus "great care" must be "taken in dealing with the infected countries" (48).

Entire countries are infected according to Bracken's rhetoric, which fits into a broader yellow peril discourse that figures Asian sexuality, via tropes of the horde, as significant threats to white civilization. At a moment of economic uncertainty over the commercial influence (and, after the Sino- and Russo-Japanese wars, military influence) of East Asia, images of a machine-like and endless labor force coincided with images of an excessive and deviant Asian sexuality; to gloss Marx, visions of an "Asiatic Mode of Production" coincided with visions of excessive Asiatic reproduction. The mass migration of Asiatic populations coincides with their apparent interspecies reproduction of microbes through intimate contacts: "kissing, nose rubbing, cohabitation," and "reception of the secretions from lepers" are figured as important modes of disease communication (105–6). Nayan Shah explains that beginning in the 1870s, Chinese immigrants became a special target of xenophobic health policies and labor activism in California; furthermore, "since heredity and sexual contact were considered the source of disease transmission, fears of leprosy mixed with fantasies of miscegenation."[20] Zachary Gussow expands on this rhetorical connection between sexuality and contagion:

> The threat of Chinese diseases became a favorite theme in arousing American fears and hostility and in emphasizing the terrible danger that Chinese immigration posed to US civilization. Chinese diseases were not perceived to be diseases in the ordinary sense. They were portrayed as more potent and, moreover, incurable. . . . By leaving diseases nameless, or even when calling them "the incurable . . . Asian scrofula," journalists and orators rendered them more mysterious and more terrifying. [Congressional testimony claimed] their "touch is pollution." . . . With their supposed addiction to opium and delight in perverse sexual habits, their continued presence was seen as poisoning America's bloodstream through the large numbers of prostitutes entering the country.[21]

Visualizing Debility: A. W. Hitt's Tropical-Medical Photography

To further understand the intimate sense of racial untouchability attributed to Hansen's disease with the expansion of the Hawaiian settler project, it is useful to shift scale and think through how visions of the bodily transitions of Hansen's disease related public fears through written and photographic representation. Let me first briefly summarize current medical understanding of the disease: *Mycobacterium leprae*, a pathogen that is likely transmitted through nasal droplets in close and prolonged bodily contact, is a weakly contagious agent. It cannot live outside of the bodies of a few vulnerable species (mainly humans and armadillos). While many individuals have strong immunity to the bacterium, those lacking it experience damage to the skin, peripheral nerves, and upper respiratory system, primarily through the defensive action of macrophages (key immune cells) that form granulomas, ulcers where inflammatory response blockades the bacteria. As such, patients often experience disease progression through numbing of nerves in the extremities, inflammation, and granulomas, which, in advanced cases, can permanently impact the capabilities and form of hands or feet. The bones of digits may be absorbed, which contributes to the myth that the disease causes extremities to fall off. Diagnosis of Hansen's has long been divided into two main forms, the "tuberculoid" (which emphasizes the presence of a small number of granulomas), and the "lepromatous" (emphasizing multiple infections with a greater number of granulomas and systemic impacts on the nervous system).

However, these distinctions are of degree rather than kind, and most patients are diagnosed with a combined, "borderline" presentation.

Even this brief synopsis of the physiology of Hansen's disease posits the bacillus and the immune system in a drama of interspecies battle. However, it is possible to modify this description, foregrounding the interactivity of microbes moving within and across the skin-bound human body, as well as the spectrum of resulting immune balances that confound divisions between the diagnostic categories. *M. leprae* lives primarily housed in or near the extremities depending on the immune response, allowing in adaptive cases (i.e., cases of disease where the bacillus successfully adapts to the body of a human host) for physical transmission of bacteria between proximate bodies and shelter for the bacillus from the stronger macrophagic response within tissues and fluids deeper beyond the skin. The limb impairment, facial transitions, skin blemishes, and nerve damage that have come to define visual images of Hansen's disease, then, are signs of an adapted macrophagic response, the balance struck with bacteria that transit across human bodies with differing genetic vulnerabilities and immune antibodies. Thus despite long-standing etiological descriptions that view the external lesion as the active site of bacterial transmission and, thus, medical justification for untouchability, the transitioning extremity is itself the articulation of immune negotiation in the fleshy layers beneath the skin, where the human body proves to be more plastic than the microbe itself. In contrast to depictions of Hansen's as a parasitic disease speeding along routes of transpacific trade, *M. leprae* is actually a poorly adapted traveler within the skin of the human. It is the very slowness of the bacterium's incubation and progression in the body, as well as its inability to reproduce in laboratory cell culture, that created much of the uncertainty around its cause and form of transmission.

Such a characterization of Hansen's disease correlates to its low prioritization today within medical research and global health agendas despite the fact that many thousands of the world's poor continue to experience its debilitating effects without access to adequate antibiotic treatment. Yet in the era of Hawaiian annexation, prior to the current treatments, the bodily transitions attributed to *Bacillus leprae* were rendered sensational and viral, living embodiments of the apparent degeneration of human form as empire perilously entangled bodies across races and borders.

The 1902 surgeon general's report discussed above includes a final, striking set of documentation that localizes the animate connections of bacillus, macrophage, skin, nerve, and environment: a set of diagnostic lantern slides submitted by the globe-trotting US physician Addison Winter Hitt. Hitt had studied the disease as a physician visiting leprosaria in British India in the 1880s. During the early 1890s, he traveled the circuit of medical conferences across the United States as an expert on Hansen's disease. Most often the proceedings began by mentioning his use of a stereopticon projector—a magic lantern—to display flickering, ghostly black-and-white images of Hansen's patients for conference attendees.[22] Hitt also helped convince key policymakers and physicians of the need for strict screening and quarantine measures. In the 1890s, he shared his images and theories about the disease with medical audiences in Chicago, New York, Kentucky, Tennessee, and Alabama, becoming an oft-cited public health advocate in medical and policy literature. He suggested that Britain had failed to stamp out a horrific health threat, possibly expanding the number of cases by the millions via unsanitary campaigns for vaccination. The United States, if not careful about immigration and new possessions such as Hawai'i and the Philippines, could follow suit.

Hitt's submission to the surgeon general includes a set of slides illustrating the impacts of Hansen's on the human body, including a number composed during his 1894 trip to the Mungeli Leper Asylum in India. The others come from Samuel Patton Impey's *A Handbook on Leprosy*, a widely circulated book based on clinical research in South Africa's Cape Colony. These are diagnostic photos and lithographic illustrations that serve the immediate purpose of documenting the stages of progression of the disease; however, in Hitt's public use of them, the images take on a political character, working to link the vulnerability of bodily form across colonial space. In the face of a possible Hansen's disease outbreak, they communicate what Sander Gilman calls "the sense of dissolution" contained in "the image of the disease anthropomorphized."[23]

The beast and the zombie are the gothic figures that give power to these images; they promote fear of a moderately contagious disease by mining the divisions between human/animal and living/dead. The first image, which Hitt uses to depict skin blistering and its effects on facial form, includes the caption "Illustrating 'leonine aspect'" (figure 1.1). The reference is to the common association in both modern and medieval medical literature on leprosy between the face of the patient and that of a lion.[24] This portrait from Impey's South African images centers the fleshy contours of the face, prominently revealing the wrinkled forehead and the pustules surrounding the patient's oral cavity, the infection pathway so feared by Alfred Mouritz. The contrast of the "simply

Figure 1.1
Illustrating "leonine aspect," from *A Report Relating to the Origin and Prevalence of Leprosy in the United States* (1902).

frightful" face with the straight lines of the well-kept Western suit visually produce the hybrid body that Hitt calls a "walking sepulcher" (96).

Further images in the series taken in India figure the body of the Hansen's patient as inhuman or animalized. Several images focus on the crippled hands and either rolled or widened eyes of women patients dressed in cloth, with head scarves and bangles prominently marking national difference. One image, which likely demonstrates an incorrect diagnosis of leprosy, highlights the distention of legs that Hitt associates with the elephant (figure 1.2). Writing on cases of Hansen's disease and "elephantiasis" (now called lymphatic filariasis, a parasitic disease impacting extremities and often confused with

Figure 1.2 Depiction of elephantiasis, from *A Report Relating to the Origin and Prevalence of Leprosy in the United States* (1902).

Hansen's), Hitt wonders in a Kiplingesque turn at the animalization of the colonized: "It seems strange that these two powerful and destructive diseases should be in appearance so much like the two most powerful and destructive animals of the country in which they prevail." He goes on to relate what he claims is an indigenous theory of disease that associates elephant bathing ponds with the swelling, fluid accumulation, and tissue thickening of filariasis (763).

Adapting a common formal element of medical photography of Hansen's disease, Hitt foregrounds the centered posing of the hands of the patient in order to demonstrate the progression of disease through ulcers and bone absorption. Several of the images are striking in depicting the direct gaze and open stance of patients who appear unfazed by stigmas attached to these marks of difference. However, as their resistant poses circulate beyond the diagnostic encounter into the racialized debates over Hawaiian annexation, they are subtly sorted into menacing and victimized subjects. Drawing on Rosemarie Garland Thomson's distinction between the gaze and the stare, I suggest that it is possible to analyze how such poses draw affective connections between race and debility. For Garland Thomson, the photographic gaze comprehends otherness on the skin-bound level of the body, while the stare stigmatizes debility by focusing the eye on the specific site of bodily difference.[25] In Hitt's images, race operates on the level of the gaze, wherein the viewer takes in the skin tone, facial features, clothing, and background as racial clues. The localization of debility, on the other hand, occurs within conventional posings of immobilized skin, face, and limb surfaces to the center of the frame. In posing the debilitated extremities to the camera's line of sight, the punctum that attracts the stare is centered and takes on added significance in relation to racial cues in the rest of the frame. This plays out in the contrast Hitt's series displays between the apparently aggressive stances of Indian patients—where the posed hands appear to grasp toward the viewer—and the prone images of white subjects from Impey's Cape Colony. One image displays four Indian men and one woman posing by holding out their crippled hands to the viewer (figure 1.3). The turban-clad man in the center of the picture and the woman to his left hold the pose of boxers, revealing the advanced state of absorption of the digits. Unlike much colonial Indian portraiture, which often strictly polices gender boundaries by contrasting men's and women's dress and poses (figuring the woman as carrier of tradition), disease here unites patients in a masculinized pose against the viewer.[26] Other images in the series similarly construct the viewer within the grasp of men wearing turbans or sitting on

Figure 1.3 Hansen's disease patients with posed hands, from *A Report Relating to the Origin and Prevalence of Leprosy in the United States* (1902).

haunches whose hands extend toward the viewer. These images of the invading body of the colonized contrast with those of two white patients who almost appear to flinch in response (figures 1.4 and 1.5). One holds his hands up, prone and slanting away from the camera, and another displays hands clutching at his neck as if he is being strangled by the disease itself. Hitt's images thus subtly display the threat of transpacific contagion to an endangered white masculinity.

Hitt's work reflects a combination of several of the prominent understandings of Hansen's disease—a situation that was not unusual given the contested status of germ theory at the turn of the century. While it does acknowledge the bacterial cause, Hitt's submission to the report also incorporates ideas from zymotic and miasma theories, suggesting chemical agents produced by fermentation or atmospheric processes contributed to disease in those weakened by diet or other contextual factors.[27] When Mouritz wrote of "an atmosphere of leprosy," he reflected the continued influence of zymotic and miasma theories that attached ideas of purity and pollution to the air and attempted to contain them through the surveillance of putrid smells and the opening of crowded spaces of mass urban society to circulation. Thus the restructuring of colonial space and the reform of hygienic practices were consistent with both germ theory and its predecessors. Although Hitt understood *Bacillus leprae* to be

Figure 1.4 Prone patient, from *A Report Relating to the Origin and Prevalence of Leprosy in the United States* (1902).

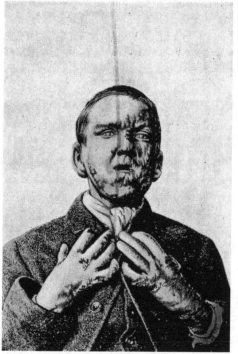

Figure 1.5 Image evoking choking, from *A Report Relating to the Origin and Prevalence of Leprosy in the United States* (1902).

the main causative agent of the disease, he also wrote in a medical journal that "a hot, damp climate or a cold, moist climate acts as an important factor in the causation of leprosy." And on the theory that specific animal-based foods contributed to Hansen's disease transmission, Hitt concluded that coastal Indians often ate "putrid" fish that would lead to skin diseases, allowing transmission of the bacteria.[28] Hitt also included two nonportrait images, one of a fern tree and another of an Indian barber, to demonstrate that humidity and poor sanitation might transmit bacteria. He presents the failure of British control measures by invoking the common theory that vaccination was causing a Hansen's disease epidemic. Reporting that schoolchildren are vaccinated directly from the lymph of infected patients, Hitt warns, "From 3,000,000 to 6,000,000 children are vaccinated in this way every year in India. Who can tell what the result will be?" (87). His focus on tropical climate, Indian dirt, and the success of quarantine in medieval Europe leads him to support quarantine aimed at "next-door neighbors" like Mexico and Cuba, possessions like Hawai'i, and ultimately all sources of travelers and immigrants to the United States (88). Hitt views India as an example of failed imperial administration from which the United States must take lessons in the Pacific.

The images presented to the surgeon general—the same ones projected by magic lantern in Hitt's live lectures—fit into what Garland Thomson calls an exoticizing visual rhetoric: one that operates around the management of distance between viewer and the object of representation. The images engage in the forms of ordering common to the racializations of South Asians in the expositions and freak shows of the era, stressing the corporeal difference and ghastly exoticism of the brown body.[29] At the same time, they draw on an archive of transcolonial visual representation that uses disease to link the racial difference and degeneration of natives from a variety of sites of tropical medical practice. Thus even though exoticized, medical portraiture aimed to incorporate the leprous body into contained racial and epidemiological knowledge, turning the racialized body experiencing microbial hybridization into a candidate for medical modernization. The exoticizing visual rhetoric was one of several modes of visual representation that contributed to "institutionalizing, segregating, and medicalizing people with disabilities." In Hitt's case, it contributed to policymakers' increased regulation of migration, travel, and trade in the form of port quarantines, as well as a series of proposed national laws to segregate Hansen's patients. Such a proposal finally passed in 1917, which led to the launch of the Carville National Leprosarium in Louisiana in 1921. The production of a medical gaze through nineteenth-century photography, argues Garland Thomson, was a key turning point in the establishment of such medical and legal institutions of quarantine.[30]

In testimony before the Senate Subcommittee on the Pacific Islands and Puerto Rico, Hitt appeared as the sole medical authority recommending nationalization of Hawai'i's Moloka'i settlements; notably, Hitt's comments carried more weight than a petition of Moloka'i settlement residents protesting nationalization of the institution.[31] Hitt's work situates Asia-Pacific racialization in the interspecies and transborder contact zones of microbial disease, and it takes its aesthetic form from the interpretation of bodily transitions as forms of degeneration or animalization. Meanwhile, white bodies are rendered as neutral victims of disease in this photographic gaze. Thus, even as he made local arguments about cultural practices spreading disease, Hitt could understand images of Chinese, South African, and South Asian patients to interchangeably represent what he encountered in colonial India and what US settlers faced in Hawai'i. Hitt's images painted Hansen's patients as embodied death, zombies, and species hybrids produced out of transcolonial contact zones.

If the photographic medium was taken in the early twentieth century as an unmediated representation of reality, this was because Western viewers had normalized the viewing position of a racially unmarked spectator following on a history of colonial photography.[32] Yet Hitt's use of lantern projection at live presentations not only allowed the use of photography and lithographic illustration to construct realistic medical knowledge of a feared disease, but worked to sensationalize the disease by presenting its most disfiguring aspects to medical audiences, accentuated by the flickering and shadowy display of the early projector and the high contrast of the form of image reproduction. Viewing such images today, it is easy to forget that these were diagnostic images for which patients were subjected to various degrees of force or even abuse in the colonial medical encounter. Diagnostic photos were often a standard part of the patient intake process. In the Hawai'i State Archives, records of unnamed patients of the Kalihi Leprosy Hospital include such intake pictures. These pictures are conventionally staged around an impairment that is posed at the center of the image, as in most of Hitt's slides. Of these many archival images that require various stages of patient undress and awkward or even eroticized poses (including of very young boys), one photograph I viewed most clearly glimpses the constructedness of the diagnostic pose. A boy of perhaps ten years stands topless with left pant leg raised and left arm crossing his chest, both limbs displaying granulomas. The right arm hangs to his waist. However, the slow exposure of the film reveals that the photographing doctor likely was moving quickly through shots and may have begun the exposure early, when another subject remained in the frame. There is a ghostly outline of what appears to be the prior patient, eyes and hair visible along with a bandaged hand held to the waist. This apparition of a hand is then covered by the loincloth of the present patient, making it appear as if the injured hand of the prior patient transits across time, reaching around the young boy to casually tuck into his pants. It is possible to read this image as revealing medical photography's own process for individuating and medicalizing patients, segregating them from their everyday intimacies among other patients and caregivers and individualizing impairment. Here the serially produced intake photo itself is a product of the force exacted on patients by the spatial regime of Hansen's disease segregation.

Dependency and the Sentiment for Segregation

In annexation-era discourses on Hawaiian vulnerability to Hansen's disease, contagion buttressed settler claims to sovereignty. If the open push for US imperial expansion in the 1890s often highlighted an anxiety about the physical fitness of whites for tropical life, discourse on Hansen's disease in Hawai'i suggested that to survive global interconnection, the Hawaiian would need to accept the outside help of US medical modernization to adapt to the risks of interspecies contact.[33] Mouritz followed his diagnosis that "an atmosphere of leprosy clings to the unfortunate Hawaiian" with an assessment of Hawai'i's dependency on settler rule: "Our duty is plain, we must stand shoulder to shoulder with the Hawaiian, brace him up, and support him until he can stand alone, like other races."[34] Invoking the Hawaiian as a racial type defined by debility and requiring the paternalistic brace of colonial medical institutions, Mouritz captures the linkage of moral failure, contagion, and quarantine that defined the response of the settler press and white business interests to Hansen's disease in Hawai'i from the 1870s through the 1910s. Rather than evidence of a political project that used illness to justify the usurpation of Hawaiian sovereignty, this public health agenda was abstracted as a form of humanitarianism. Walter Gibson, the president of the Board of Health and an editor of two Hawaiian newspapers, made clear in his 1873 comments the relationship between Hawaiian difference, enforced segregation, and the universal humanitarianism of medicine: "The horror of this living death has no terror for Hawaiians, and therefore they have need more than any other people of a coercive segregation of those having contagious diseases. Some people consider this enforced isolation as a violence to personal rights. It is so, no doubt, but a violence in behalf of human welfare."[35] As Maile Arvin demonstrates, such invocations of the human helped to justify colonial settlement and to envision the potential whitening of the figure of the Polynesian.[36]

Thus the racialization of *M. leprae* at the turn of the twentieth century contributed directly to a wide architecture of state intervention that went beyond its ostensibly medical sphere of control. This was widely circulated in the sensational literature of colonial encounter published by US and European authors.[37] In this section, I turn to Jack London's Pacific narratives, which constructed Asia-Pacific racial forms in relation to a tropical-medical modernity that sentimentalized colonial dependency in the face of contagious diseases like Hansen's. Rather than turning away from the stigmatized leprous body, racialized subjects of quarantine were both rendered as underdog heroes and subjected to a settler historicity that normalized able-bodied whiteness against the mutating and deathly Hansen's patient. As such, they posited whiteness as the future of tropical modernity and the native Hawaiian as inevitably engulfed by the transformations of transpacific contact. In the process, depictions of Hawaiian patients sentimentalized the institutional dependency of the leper figure, ascribing disability to their bodies in the process.[38] The rendering of the reformed leper as dependent on public goods plays a key role in disavowing the initial violence involved in the forms of settlement and expropriation—violence that itself destroys preexisting interdependencies, other relations to sociality, and forms of life among those rendered disabled.

To illustrate, let me begin with a scene of interdependence: in Hemel-en-Aarde (the Dutch "leper farm" at Cape Colony, South Africa, which preceded Impey's Robben Island settlement), a traveler views a scene of two men. One, having no working arms, walks supporting the other, a man with no legs, on his back. The man with no legs drops peas onto the ground at intervals as the one walking stomps them into the dirt.

There is no dearth of such orientalist tales of the interdependence of the leper figure of the late nineteenth century. The scene above, related in Charles Warren Stoddard's 1885 narrative of his journey to Moloka'i, is taken from a widely cited midcentury account by Bishop Hallbeck.[39] The image of the two interdependent bodies operates by producing wonder at the survival and adaptation of the exoticized and institutionalized Hansen's patient. It accomplishes this mirroring and differentiation of the able and disabled body by dismembering the body, and then by doubling this dismembered body to produce a simulacrum of the complete, able-bodied human. In a text that vociferously upholds the Christian duty of outsiders to care for Hawaiian patients, the doubling of the dismembered body is the textual site at which an orientalist rendering of the diseased body can co-opt the real and often necessary interdependence of individuals living with *Mycobacterium leprae* into a spectacle reinforcing colonial dependency and segregation. Imperial discourse appears preoccupied with the monstrosity and primitivity of bodily interdependence; it simultaneously normalizes the dependence of Asian and Pacific Islander bodies upon missionary tropical medicine as a form of progress.

Like Stoddard, Jack London was one of the few mainland visitors granted entrance to the Moloka'i settlements. Unlike Stoddard, however, London was invited by the Board of Health expressly to defend the colony against its sensational portrayal in the media. In his 1908 narrative, "The Lepers of Molokai," London does just that. Narrating the pleasures of hunting and horse racing during his July 4 visit, London depicts the happiness—based on a sense of community—of those at Moloka'i. He does so not simply to counter stereotyped images of patients as zombies, but also to justify enhanced surveillance and quarantine, demonstrating the enhanced colonial enforcement of segregation rather than its timeless character: "That a leper is unclean . . . should be insisted upon; and the segregation of lepers, from what little is known of the disease, should be rigidly maintained. On the other hand, the awful horror with which the leper has been regarded in the past, and the frightful treatment he has received, have been unnecessary and cruel."

The difference from Stoddard's construction is the absence of a monstrous or exoticized depiction of the leper figure. Nonetheless, he portrays in the patient's institutionalization a proper medical citizenship. London's idealization of patients' institutional dependency is unexpected—it reads more like paternalistic New York Progressivism than the masculinist California socialism that for London involved expressions of disdain at Asian immigrants. Segments of the California labor movement such as the Workingmen's Party had championed the racialist association of leprosy and Chineseness to justify immigrant exclusion and minority marginalization in San Francisco, for example, in the late nineteenth century.[40] Colleen Lye recounts the ways in which London represented the Chinese as a "yellow peril"—a despotic specter of the coming capitalist modernity—against the primitive Pacific Islander destined to be overcome by the rise of a homogenizing Asia. The Anglo-American was caught in between these two poles of hypermodernity and primitivity in the Pacific world. According to Lye, however, once the Chinese enter the Hawaiian setting in London's writing, they are subjected to an "Edenic Polynesian effect," losing their overt villainy in the primitivized Pacific setting.[41] Arvin, relating London's fascination with the Hansen's patient to scientific racialism, notes that London found in the so-called leper a type of hidden disease that marked the indigeneity that hid behind the apparent near-whiteness of the Hawaiian.[42] In both of these readings, the Hawaiian setting is distanced from London's depictions of the teeming masses of continental Asia, which, in London's essay "The Yellow Peril," will provide the labor power for the hyperrational production schemes of emerging Japanese empire. Furthermore, the specific setting of quarantine brings the Chinese and Hawaiian outside of the capitalist world that structures London's Pacific Rim racial thought. Unlike the Chinese coolie figure demeaned in Asian exclusionist fiction written at the turn of the twentieth century, the Moloka'i narrative identifies the Portuguese, Hawaiians, and Chinese populating the segregated leprosy settlements — people who remain outside Pacific labor markets — as proper objects of sympathy.

Unlike many of his other works, London's essay is not explicitly anticapitalist or anti-imperialist. Dependency on US charity seems vital to the advancement of the medical treatment of the colonists, and proper given the controlling US presence on the islands. London does, however, retain an object of racialist scorn as foil to the otherwise noble colony residents. A black colony resident who has been screened and declared disease free becomes the focus of several paragraphs of London's narrative. Despite his supposedly clean bill of health, the man "preferred Molokai" and fought expulsion from the settlements. London depicts the man's stated desire to remain at Kalaupapa as a "game" intended to ensure for himself the steady flow of government resources available for colony residents — to ensure his continued parasitic dependency on the state while pathologically remaining outside of what London saw as the proper forms of individualistic interdependence at Kalaupapa. Black freedom and black dependency are both figured as threatening, and London muses jovially at the fact that the man is ultimately returned to Honolulu. Although this discussion is ostensibly contained to a single black resident, such criticisms were commonly leveled against native Hawaiian kōkua who were suspected of wishing to stay on Moloka'i in order to receive free food and other state assistance. (This despite the fact that kōkua were maintained in order to reduce labor costs.) Thus the dependency of Hawaiians and Chinese that London naturalizes comes into a disavowed narrative intimacy with what he sees as the pathological dependency of blackness. There is thus in London's reporting a tendency to typologize particular relationships to dependency as features of racialized groups. London's writing supports stereotypes of blacks as unproductive, unruly, and resistant to tropical disease, and reveals an ambivalence toward a settler project that needs to provide public goods to those he cannot verify as deserving. On the one hand, taking on institutional dependents seems to be the responsibility of the United States given its claim to the islands; on the other, it allows some groups to steal from the state at the expense of the white worker.

These rhetorical strategies also allow London to situate the most important source of Hansen's disease elsewhere, the failure of that colonial power whose example most worried A. W. Hitt: "There are half a million lepers, not segregated, in India alone." The white Hawaiian business interests whose plantation economy was ecologically associated with speeding the transit of Hansen's disease on the islands escape blame as London idealizes the Board of Health's efforts at quarantine; meanwhile, it is key that the Indians with the disease under British rule are "not segregated," threatening the real global pandemic feared by those who sensationalized Molokaʻi.[43] Addressing Rockefeller, Carnegie, and other robber baron philanthropists who were funding public health initiatives and research in tropical medicine, London wryly concludes that Molokaʻi is "the place for your money, you philanthropists." The essay ultimately asserts the humanity of patients to justify their enforced segregation and dependency. London's humanization of settlement residents can only operate in a context in which colonized dependents are carcerally enclosed in a space outside of transpacific capitalist production.

"The Lepers of Molokai" transforms the scene of dependency in order to incorporate the patient into US imperial designs, justified with reference to a medicalizing humanitarianism. Yet even though London attempted to combat gothic and other sensational images of the disease in his 1908 piece, such images actually open up a variety of possibilities for representation that London was yet to explore. The gothic is not simply a mode of exclusion. It is subject to excessive meanings that invest the body with a desiring gaze. Stoddard wrote of the "childlike" Hawaiians (116), "They would smile in their last breath; for of all nations on the face of the globe the Hawaiian is perhaps the most amiable and the most ingenuous. But what smiles were those that greeted us!" (55). Like London, Stoddard supported segregation, and his complex sentiment for Hawaiians—treading the border of the erotic and the phobic—played off of other groups. It was likely that "special laws" would be necessary on the mainland for the "segregation of those who have fallen victims to the most dreadful of all scourges." This is especially true of the "Chinese coolly," whose migration sows "the seeds of the plague" (121–22). Yet even as the disease was a condition in which "the living" was "well-nigh dead" (92), the debilitated spectacle of the Hansen's disease patient's body was rendered a fantastic object of desire. Consider the scene in which Stoddard's narrator is led to the bed of a patient at Molokaʻi. Here, even as the mouth of the patient provokes horror as it did to Mouritz, ulcerated flesh hides within it the fruits of desire:

> A corner of the blanket was raised cautiously: a breathing object lay beneath; a face, a human face, was turned slowly toward us—a face in which scarcely a trace of anything human remained. The dark skin was puffed out and blackened; a kind of moss, or mould, gummy and glistening, covered it; the muscles of the mouth, having contracted, laid bare the grinning teeth; the thickened tongue lay like a fig between them; the eyelids, curled tightly back, exposed the inner surface, and the protruding eyeballs, now shapeless and broken, looked not unlike bursted grapes. . . . Surely the grave knows nothing more frightful than this! (99)

If the gothic produces horror by using the abnormal to shore up the position of the normal, it is always still subject to "a vertiginous excess of meaning"; it produces varied possibilities in the monster, rather than a single position in the order of things.[44] Gregory Tomso has utilized Stoddard's description above of the patient beneath a blanket as an example of *leprophilia* given the blurring of the medical gaze with the sexual gaze. Whereas Mouritz expressed particular worry about the mouth and Hitt focused on the eyes and the facial transformations of the Hansen's patient, Stoddard's images of the tongue as fig and the eyes as grapes suggest an orientalizing desire at the very bodily sites of surveillance where Hitt witnessed the degeneration and animalization of the body. For Tomso, "the body of this leprous child is nearly edible, the ripened flesh teasing if not tempting the reader with exaggerated versions of the same erotic metaphors of fruit and flesh that Stoddard uses in his quasi-pornographic stories" set around the Pacific.[45] The gothic here affords Stoddard an expression of queer desire that otherwise cannot be made explicit— the narrative never mentions Stoddard's cohabitation with a male resident of the Kalawao settlement at Molokaʻi. Flesh that is elsewhere considered rotten emerges as ripe, suggesting that the hybridization of the human by the Hansen's bacterium produces complex sensations treading the borders of the erotic and the phobic.

Oddly, London also demonstrates this play between desire and horror at the body of the Hansen's patient. He shifts from sentiment and sympathy to an eroticized gothic rendering of Moloka'i one year later in the 1909 story about indigenous resistance to US empire, "Koolau the Leper." In "Koolau," London deploys gothic sensationalism to represent Hansen's disease, but maintains identification with the disabled Hawaiian subject, whose body becomes the site linking the power of US empire, finance capital, and the emerging orders of racial and sexual difference. The story follows the historical figure Kaluaiko'olau (Ko'olau), who in 1893 successfully avoided transportation to Moloka'i as a fugitive in the Kalalau valley, Kaua'i. Although many individuals fled the Hansen's policy and attempted to settle near other fugitives in Kalalau rather than accepting removal to Moloka'i, London focuses on a single resister, extracting Ko'olau's story from the broader political battle in which native Hawaiians with the disease attempted to establish regional settlements that would offer more autonomy.[46] Ko'olau famously killed a sheriff and outmaneuvered two groups of provisional government soldiers sent to hunt him down. In London's text, the character Koolau (I preserve London's misspellings, which lack the 'okina mark) is a figure of masculine resistance to US empire. Koolau refuses both stigmatization and a larger structure of colonial degeneration that London describes through the emergence of Hansen's disease. The story begins with his speeches, which recall an idyllic past before land enclosure destroyed Hawaiian livelihoods and disease—brought by plantation labor—devastated Hawaiian bodies. These speeches, delivered in a participatory meeting of Koolau and other fugitives of the quarantine policy, are followed by a hula celebration featuring dancing and the use of indigenous narcotics. The scene is broken by a rocket fired in the distance, which signals the arrival of the American-led Hawaiian police. The fugitives surrender, betray Koolau, or are gunned down after initially resisting arrest. Erasing Ko'olau's real-life wife and son, the narrative sees only Koolau escaping the initial attack. Eventually he too is wounded and dies alone with his gun.

Hawaiians with Hansen's disease become the site of affective excess, an erotically charged transition toward the monstrous—even as they embody resistance to imperialist logics of economic and medical dependency. London first invents a subaltern voice to critique the violence of US empire in the Pacific. He portrays Koolau as ruler of the fugitives in Kalalau, his oratory providing the perspective of native informant critiquing white settler violence. Koolau's opening speech portrays early missionaries and traders as underhanded avatars of colonial dependency and Hawaiians as their victims:

The one kind asked our permission, our gracious permission, to preach to us the word of God. The other kind asked our permission, our gracious permission, to trade with us. That was the beginning. To-day all the islands are theirs, all the land, all the cattle—everything is theirs. They that preached the word of God and they that preached the word of Rum have foregathered and become great chiefs. . . . They who had nothing have everything, and if you, or I, or any Kanaka [Hawaiian] be hungry, they sneer and say, "Well, why don't you work? There are the plantations."[47]

In "Koolau," the monoculture plantation brought by the American settler is the site of economic and biological violence against the indigenous. Figuring the eighteenth-century Hawaiian as premodern and possessing a proper pastoral relation to nature, by the end of the nineteenth century, Hawaiians have succumbed to a disease of civilization. Unlike "The Lepers of Molokai," which constructed India as the global disease threat in defense of Moloka'i as a Polynesian paradise, London returns to a yellow peril rhetoric, locating the Chinese source of the disease in labor migration paths. Koolau continues, "Because we would not work the miles of sugar-cane where once our horses pastured, they brought the Chinese slaves from over seas. And with them came the Chinese sickness—that which we suffer from and because of which they would imprison us on Molokai" (20). Like London's use of underdog masculine heroes in a variety of works, the heroic oppressed voice here protests US empire from the perspective of a class-based critique. For London, US commercial enterprises exploit natives and produce effete cultures of imperialism among the dominating classes.[48] Meanwhile, economic marginalization defining the space of subalternity becomes the basis of medical dependency via the importation of the Chinese, who were often represented as a contagious human virus.[49] Thus London's attempts to speak the voice of those dispossessed by US empire tend to masculinize the subaltern in the face of disability that produces dependency. Unlike the Chinese in London's yellow peril writings, the Hawaiian is not portrayed as a machinelike, inhuman worker because his authentic relation to nature (and thus his vulnerability to extinction under capitalism) prevents his corruption. The Hawaiian instead presents a dying breed of people whose proper relation to a dying nature—London's melancholic object—has been sidelined by transpacific capital and Asiatic despotism.

After Koolau's opening speeches, London's effort at masculinizing Hawaiian resistance is complicated by an eroticized, imperialist gaze upon the leprous body during a hula scene. At first, Hansen's disease desexualizes the bodies on display to the narrator. As Koolau delivers his opening oratory, the narrator describes the scene as troubling the categories of sex: "They were creatures who had once been men and women. But they were men and women no longer. They were monsters—in face and form grotesque caricatures of everything human" (19). Hawaiian fugitives with Hansen's disease appear "barbaric," "apelike" "human wreckage" (22, 20). But as in Stoddard's writing, the horror invested in the animalized and debilitated body sets the stage for queer desire. The narration shifts to describing the cultural life of the fugitives through an orientalist gaze, removing Koolau from the scene. Hula dancing and ingestion of the narcotic *ti* plant are the sole diversions from the harsh conditions of fugitive life. Here, the narrator argues that only a drug-induced hallucination can produce the paradoxical forgetting that enables the fugitives to think of themselves as human (sexed) again: under narcotic influence, "they forgot that they had once been men and women, for they were men and women once more" (22). The drug induces a liberatory historical amnesia, returning to the fugitives the gendered sensibility of being human, if not the predisease bodies to which normative gendering attaches.

Yet as the narrator's gaze rests on the dancers' bodies, the gothic disfiguration of the female body shows in detail how empire is "travestying love": next to Koolau's protégé Kiloliana, in whose movements "love danced," "was a woman whose heavy hip and generous breast gave the lie to her disease-corroded face." In this "dance of the living dead" in which bodies "loved and longed," the female is singled out for her compromised beauty. As Kiloliana leaves the scene, another figure appears dancing on the mat, this time with different impairments. Her "face was beautiful and unmarred," but she had "twisted arms . . . marked by the disease's ravage." The scene closes with an image of these "two idiots" who "danced apart, grotesque, fantastic, travestying love as they themselves had been travestied by life." Ending with two women—or, in London's formulation, women-turned-monsters—"travestying love," London dismembers and doubles the leper figure, following the schema of Stoddard's pea-planting scene. The doubling of the disabled body produces a simulacrum of the sexed body, eliciting the narrator's desiring gaze. London renders the leper figure monstrous through a queering of disabled sexuality. Disability and sexual nonnormativity signal dehumanization in what the narrator calls a "dance of the living dead" (22).

In the excessive signification of the gothic, the possible fears of the narrator multiply. For the mind to fill with "maggots crawling of memory and desire" (i.e., for the smooth functioning of heteronormative desire), must the disabled woman—like the mythic pea-planters at Hemel-en-Aarde—be doubled? Is the true "travesty" of love the possibility of same-sex desire between the women? Is the possibility of such desires a result of the expropriation of Hawaiians' land and bodies, a reflection of the symbolic castration of the (masculinized) indigene by the pestilent interspecies intimacies of empire? Does the true horror emerge from the possibility that the imperial gaze could produce (cross-racial) erotic desire even as the colonized body transitions into a hybrid state, in the process rendered dependent by the US colonizer and his Asiatic labor force?

These questions go unanswered as the colony is invaded and as most inhabitants are killed or imprisoned by US soldiers; the fugitives' refusal of dependency on segregated imperial health institutions is the occasion for colonial violence, as the plantation interests behind the segregation policy draw on the heavy hand of military policing. Here the narrative returns to the perspective of the native informant, Koolau. As the remaining survivor, he is left alone to die in the forest with his rifle and with the satisfaction of having killed white Americans. There is no need for sexuality or even for sociality in his lonesome end—he simply dreams of his "early manhood," desiring death. Like many of London's lone male heroes, Koolau is triumphant but dies alone, unable to reproduce his naturalized masculinity in the face of hypermodern capital. Having been betrayed by several of his fellow fugitives, the only possible masculine identifications—with "brave" American soldiers—are foreclosed by the soldiers' association with an empire that values the Hawaiian man's dead body, but not his independence: "Because he had caught the sickness, he was worth a thousand dollars [ransom]—but not to himself. It was his worthless carcass, rotten with disease or dead from a bursting shell, that was worth all that money" (28). In this lonesome ending, Koolau's body serves as what disability theorists David Mitchell and Sharon Snyder have termed a "narrative prosthesis": disability becomes a metaphor for social disempowerment, casting aside the problems of living with impairment or with the social relations that emerge from biological precarity.[50] These problems—of producing interdependencies in the interest of the community of the disabled—ultimately have no place in London's vision of masculinity. The fugitives are ultimately killed or forced into hierarchic dependencies of empire.

The questions raised by the hula scene—and cast aside by the resolution—deserve attention. Female homosexuality is particularly unusual in representations of the disease, and London's writing rarely foregrounds women as desiring figures. I see the unusual scene as a result of two interrelated gendering processes: the first feminizes the majority of Hawaiian fugitives in order to isolate Koolau's masculinity as transcendent, and the second feminizes the social modes of interdependence that bind the fugitives. Given the centrality of the dependency dialectic in London's Molokaʻi writing, it appears that sexuality (as female homosexuality) emerges as an uncontainable site of embodied interdependence in the text. The idea that the normative wholeness of the able body is rendered unnecessary through a social erotics is anathema to London's vision of independence, and in his response sexuality itself is ultimately feminized. As the pathologically sexed women "travesty love," Koolau is conspicuously absent from the single scene in which the narrative shifts away from his viewpoint.

In the context of disability, community interdependence is essential for the reproduction of London's atomized vision of culture (here symbolized by the hula). London attempts to feminize sexuality as a form of dependency, in contrast to the masculine hero who must always emerge as independent. But this ultimately means that Hawaiian culture—the mark of Hawaiian difference—is feminized at the moment London wants to assert Hawaiian masculinity and independence.[51] The primitivist depiction of Hawaiians works against the legibility of individualist masculinity that London attempts to position as transcendent. As Denise Ferreira da Silva writes, "culture"—in the sense of a bounded set of identifying traditions of a population—is the key mark of difference within the racial politics of modern coloniality, and a basis of both the oppression of the darker nations and the production of their anticolonial subjectivities as seemingly transparent "voices."[52] London's use of a cookie-cutter lone male hero runs into the problem of having to signify subaltern difference in a colonized setting:[53] how can Koolau embody Hawaiian resistance if his dying body alone cannot reproduce the Hawaiian difference that underwrites the view of Hawaiians as innocent victims of the plague of empire? London needs to return to the voice of the oppressed through a figure he calls "king" of the fugitives—the sovereign. He cannot allow the hula or the uninfected caregiver to speak an alternative set of intimacies or desires that would acknowledge that living with disease-causing bacteria could queer anti-imperial politics; he instead subjects difference to a controlling imperialist gaze. Narrative prosthesis thus allows his vision of anticolonial resistance to avoid the lived issues of disability, sexuality, and community, while universalizing masculine individualism.

I see the contradictions between the two London texts as resolvable only in a context in which literature attempts to capture Hansen's disease through the technology of race, by attaching race to debility. As Albert Memmi explains, dependency relations suffuse colonial projects and reflect not only the dependency of the colonized upon the colonizer, but also of the colonizer upon the colonized.[54] London's depiction of Koolau is not just a vision of Hawaiian resistance but a figure who can be transmuted into a settler masculinity that London sees as endangered by Pacific border-crossings. Idealizing Hawaiian resistance requires signifying Hawaiian cultural particularities while ultimately dismissing them in the service of either a universalizing masculinism or an imperialist and paternalistic reform discourse. The Pacific Islander as a racial type stands primitivized and dependent under imperialism, particularly given that she is threatened by another racial form—the Asiatic—who, embodying the highly rationalized labor formations of what Marx called the Asiatic Mode of Production, is represented as excessively productive and reproductive. The Asiatic is the virus of transpacific production, unleashed by the entanglements of US-led capital that began to undo Europe's grasp on global finance in the late nineteenth century. London's texts thus produce heroes by engaging in a racial typing common to naturalist discourse. In fact, in examining London's depictions of Asia and the Pacific generally, Colleen Lye claims, "American naturalism represents a failed critique of capitalism. . . . The evidence of this lies in its tendency toward racialization, or the reification of social relations onto physiological forms, or types."[55] I would add that naturalism of London's brand reifies social relations by attaching racialized groups to particular types of biological transition (disease) and gendered dependency (institutionalization).

Engendering an Indigenous Critique of Segregation

A photograph published in 1906 scrambles London's attempt to extricate a lone male hero from the racialized and feminized interdependencies of Hawaiian collective life. In a stunning homage to her late husband Koʻolau, Piʻilani Kaluaikoʻolau dons the same clothing of his that she had worn in exile in the Kalalau valley (figure 1.6). She holds a shotgun in a re-creation of Koʻolau's surveillance of white soldiers of the provisional government. Embodying the ideal of the lone male hero, Piʻilani's act of cross-dressing signifies a crisis for London's production of a heteronormative masculine hero of resistance to quarantine.

The image was published along with her memoir of the events of her family's exile to Kalalau. Piʻilani's story of escape from quarantine authorities

Figure 1.6 Portrait of Piʻilani Kaluaikoʻolau, originally published in *Kaluaikoʻolau!* (Honolulu: Kahikina Kelekona / John Sheldon, 1906).

can be read as a critique of both the violence of the imperial quarantine imposed by white outsiders and, implicitly, the gendered dependency on Christian heteronormative family structures imposed on the islands. *The True Story of Kaluaikoʻolau, as Told by His Wife, Piʻilani* employs a devotional narrative to offer a counterpoint to sensational narratives of the conflicts over segregation orders in Kalalau at the moment of annexation. Piʻilani's narrative of Koʻolau's speeches against the isolation order begin with an expression of the couple's loyalties to each other, which trump any expected loyalties to the settler state. In idealizing the marriage and the family (Koʻolau and Piʻilani brought their son into exile), familial interdependence displaces the hierarchical dependence on colonial public health. However, despite the heavy-handed heteronormative mode of the narration, the story oddly presents marriage as itself a different prison. As Koʻolau states, repeatedly, "They think that I shall be imprisoned by the sickness, but marriage is the only thing that keeps this body as a prisoner."[56] The narrative presents Koʻolau as heroically and selflessly taking his family into exile, to live humbly on the plants, fish, and shrimp of the Kalalau valley and its rivers. Piʻilani cross-dresses in her husband's clothing, joining him on a constant lookout for police. Yet this fugitive bond—this bond that excessively positions devotional Piʻilani as a second, masculinized embodiment of Koʻolau—is presented as a sort of entrapment. When friends encounter the family in hiding, long after the state's pursuit ends, Koʻolau's decision to stay hidden in the "deep gloom" of the mountain leads to a life of isolation akin to Koʻolau's description of enforced quarantine.[57] Koʻolau and the couple's son die, and Piʻilani finds herself unable to leave the valley, fearful for her life and, perhaps, unable to envision rejoining society.

Piʻilani's vision of a family of fugitives leaving society and bending gender roles in the pursuit of freedom implicitly questions the freedom of various "outsides" to quarantine (domesticity, fugitive life), and furthermore questions the internalized forms of discipline that emerge from the figure of masculine resistance to empire as depicted by London. Sidelining the fantasy of escape from quarantine as well as the myth of heroic exile, Piʻilani dons the costume of the rebel but also meditates on the loss of communal life and her complex relationship as a displaced fugitive to native Hawaiian land. Despite her sense of alienation and loneliness, the narrative locates kinship both across boundaries of species in the wonder of the Kalalau valley ("this entire valley from its high cliffs to the flat terraces of earth")[58] and back to the national community she lost in exile. Piʻilani's body thus becomes a site of queer kinship that exceeds the heteronormative devotional mode through which her narrative initially seems to create moral authority.

Whereas Pi'ilani goes fugitive to maintain the form of her nuclear family, Olivia Breitha, a daughter of Portuguese immigrants diagnosed with Hansen's in 1934, recounts her separation from a fiancé as well as a traumatic segregation experience in her memoir. Her intake photo at Kalihi Hospital appears on the cover of Breitha's memoir, *Olivia: My Life of Exile in Kalaupapa* (figure 1.7). Like that of Pi'ilani, this is a defiant image of an individual subjected to public health surveillance. Displaying her hands in the conventional pose of the Hansen's medical portrait, Breitha aims her gaze directly at a medical viewer whom, in an autobiographical treatise in 1988, Breitha takes to task for the violence of incarceration and stigma.

As patient activisms developed within the expanding Kalaupapa and Kalawao settlements at Moloka'i, a variety of new possibilities emerged for reframing the history and struggles of Hawaiian patients impacted by the segregation order. This was especially the case by the 1940s, when Hawaiian Hansen's activists demanded new pharmaceutical treatments like the drug Promin that had been made available to mainland patients at the Carville Hospital. By the late 1950s, patient activisms led to the cancellation of blanket isolation policies across US-controlled territories. Yet this was not the inevitable outcome of antibiotic therapies; in fact, for a decade after the first effective use of Promin at Carville, doctors demanded even stricter segregation and institutionalization practices in order to ensure proper administration of drugs. Only after sustained patient activism at Carville and Kalaupapa demonstrated the efficacy of drug therapy and the weak justification for quarantine did public health officials embrace the end of enforced segregation for Hansen's disease.

Breitha's autobiographical narrative, *Olivia: My Life of Exile in Kalaupapa*, is straightforward in its humanist avowal of the many forms of solidarity that arise in the context of Hansen's settlements and in broader contexts beyond Moloka'i. Intimacies—from the author's multiple live-in relationships, to the relations of care and violation in medical treatment, to the stories of her parents' attempts to relocate to Moloka'i—structure the life chronicle. It is such intimate relationships, particularly with a man named John, in which mutual aid helps Olivia build a life on Kalaupapa and oppose the injustices of quarantine regulation. While intimacy is often stigmatized in the imperial discourse on Hansen's disease, Breitha makes the medical gaze itself the most obvious site of immorality and violence. After being admitted to Kalihi Hospital, Breitha recalls being subjected to a performance in what she calls the "mon-

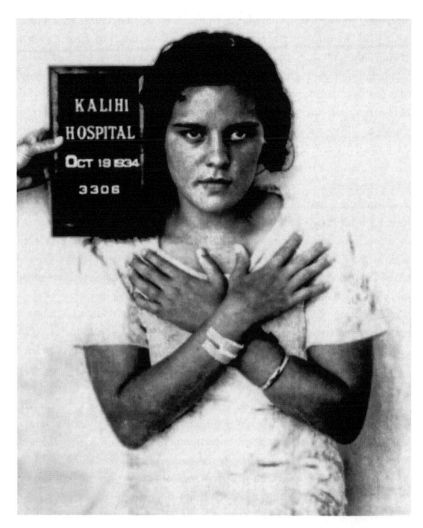

Figure 1.7 Intake photo of Olivia Breitha from Kalihi Hospital (1934), originally published in *Olivia: My Life of Exile in Kalaupapa* (1988). Courtesy Estate of Olivia Robello Breitha.

key show," in which she is paraded in front of a crowd of doctors in her underwear (which a nurse partly removes while she is onstage). Breitha writes of "being filled with shame because I was an adult, yet I had no control over what was being done to me."[59] Breitha recounts and protests the loss of freedom of movement, substandard provisions, and painful fumigation. At the end of her narrative, Breitha extends these lessons in solidarity with AIDS patients of the 1980s: "Please don't panic. Please don't do to people with AIDS what you did to us."[60] The violent institution of quarantine against HIV-positive Haitian refugees (see chapter 5) is a tragic reminder that Breitha's plea was not uniformly heeded.

A number of colonial historians of Hansen's disease working on sites ranging from Hawai'i and the Philippines to South Africa and India document both the racial discourse that constituted dark-skinned bodies as contagious vessels for leprosy and the challenges that colonized patients offered to the regimes of segregation, including incipient articulations of biomedical citizenship.[61] From this perspective, it is important to look broadly at the history of Hansen's disease segregation in the early twentieth century as a transcolonial circulation of public health knowledge, technologies for the government of species, and emergence of patient identities. Although Michelle Moran brilliantly shows how patient activisms were often divided by the racial logics separating patients at Carville and Kalaupapa, there are moments in which their interests intertwine, for example, when one Carville patient advocated reforms at Kalaupapa and when Kalaupapa patients visited Carville to learn about Promin treatments.[62] These moments of intersection build on a longer history of patient organizing on Moloka'i, which was aided by settlement residents' ability to vote and their organization of petitions and attempts to influence treatment. Refusing dehumanization by the media, religious and medical institutions, and the Hawaiian and US governments, quarantined patients forged cultural, social, and physical interdependencies as well as a sense of subjectivity in their space of confinement. They formed a Kalaupapa chapter of the Hui Aloha 'Āina (Hawaiian Patriotic League), a group organized to resist the illegal takeover of Hawai'i by the United States in 1893. This chapter added the names of some five hundred residents and nonresidents to the 1897 petition of "native Hawaiian subjects and residents" who "earnestly protest[ed] against the annexation of the said Hawaiian islands to the said United States of America in any form or shape."[63] Patients actively advocated for changes in appointments of head doctors and the introduction of palliative treatments that the Board of Health resisted. Protest played a pivotal role in improving quality of life and eventually ending the institutions of forced segregation.

Rethinking the Camp at Moloka'i

In *Homo Sacer*, Giorgio Agamben recognizes one additional transnational route that incorporates the Moloka'i settlements into the circuits of modern colonial power. Analyzing the testimonies of *Versuchpersonen*, the "human guinea pigs" subjected to vivisection in the Nazi camps, Agamben takes note of the similarities of their predicament to that of US prisoners used for infectious disease research, including death row prisoners who were subjected to experimentation in the occupied Philippines and Hawai'i in exchange for reduced sentences. The Nazis cited these precedents in their defense at the Nuremburg trials, and in response the US American judges proffered the notion of "informed consent" to distinguish between the American and the German research subjects. For Agamben, the "hypocrisy" was "obvious":

> What the well-meaning emphasis on the free will of the individual refuses to recognize here is that the concept of "voluntary consent" is simply meaningless for someone interned at Dachau, even if he or she is promised an improvement in living conditions. From this point of view, the inhumanity of the experiments in the United States and in the camps is, therefore, substantially equivalent. . . . If it was theoretically comprehensible that such experiments would not raise ethical problems for researchers and officials inside a totalitarian regime that moved in an openly biopolitical horizon, how could experiments that were, in a certain sense, analogous have been conducted in a democratic country? The only possible answer is that in both contexts the particular status of the vps [Versuchpersonen] was decisive; they were persons sentenced to death or detained in a camp, the entry to which meant the definitive exclusion from the political community. Precisely because they were lacking almost all rights and expectations that we customarily attribute to human existence, and yet were still biologically alive, they came to be situated in a limit zone between life and death, inside and outside, in which they were no longer anything but bare life . . . a life that may be killed without the commission of homicide. Like the fence of the camp, the interval between death sentence and execution delimits an extratemporal and extraterritorial threshold in which the human body is separated from its normal political status and abandoned, in a state

of exception, to the most extreme misfortunes. In such a space of exception, subjection to experimentation can, like an expiation rite, either return the human body to life (pardon and remission of penalty are, it is worth remembering, manifestations of sovereign power over life and death) or definitively consign it to the death to which it already belongs. . . . In the biopolitical horizon that characterizes modernity, the physician and the scientist move in the no-man's-land into which at one point the sovereign alone could penetrate.[64]

I began this chapter with a discussion of Alfred Mouritz, whose experiments on Molokaʻi led him to declare the mouth of the Hawaiian to be the "path of the destroyer," the infection pathway of *Bacillus leprae*. At the same time that Mouritz was head doctor at the Kalawao settlement, the Anglo-German microbiologist Eduard Arning conducted research there on Hansen's disease. Arning notoriously offered a deal to a death row inmate named Keanu, who would avoid execution and serve a life sentence if willing to become Arning's experimental subject. Keanu thus "volunteered" for the research and in 1884 was implanted with an infected granuloma from another patient. He died of Hansen's disease in 1892, though it was unclear if Arning's experiment had directly infected him or if he had contracted the infection from another source.[65]

Agamben mentions Keanu as one defense example at the Nuremburg trials. With hindsight, the record of experimentation at Molokaʻi was more extensive than this single case cited in court. As Nicholas Turse explains, Molokaʻi became known internationally as a colonial research center on Hansen's disease, and was beyond the medical mainstream in the 1880s and 1890s in the extent to which it allowed human vivisection.[66] Head physicians at the settlement, including Mouritz and George Fitch; researchers invited or sponsored by the government, including Arning; and significant international authorities on the disease, including Armauer Hansen himself, had all engaged in human experiments during this era. In addition, dozens of nonhuman animals (mainly monkeys, rabbits, and flying insects) were used to attempt to prove the bacteriological theory of transmission. In the 1880s, the American Humane Association expressed horror at a series of experiments that attempted to infect healthy children with Hansen's, even as recent anticruelty statutes safeguarded some animals from similar testing.[67] Pacific leprosaria became key sites for experimental treatments for Hansen's. With no indigenous European treatment for the disease, colonial doctors in India had appropriated the main ayurvedic treatment of chaulmoogra oil massage for experimental use.[68] By 1901, medical authorities had instituted use of the oil as the primary treatment for Hansen's in both the San Lazaro hospital in the Philippines and the Moloka'i settlements.[69] British doctors developed a notoriously painful method of injecting components of the oil, squaring the ayurvedic treatment with allopathic logics of intervention by incorporating it into an injectable compound made to interact directly with circulatory and immune systems.[70]

Agamben's analysis is compelling in its ability to connect the inflationary demand for scientific knowledge to the uneven territorialization of the sovereign power to kill. However, Agamben seems to conjoin control over life and death as the single modus operandi of sovereignty despite Foucault's distinction of a sovereign power to kill from a biopolitical power over life, which for Foucault is more insidious and operates through the subject rather than only through the body. Given the ostensibly humanitarian character of the reform of quarantine that I have documented throughout the chapter, it is useful to inquire further into Agamben's consignment of the experimental subject to the space of exclusion, to bare life. The story of Moloka'i, including its histories of experimentation, may offer other lessons for the study of empire's manifestation of power in the body.

If individuals diagnosed with Hansen's disease were persistently denied legal rights, they were also increasingly the subject of reform discourses that suggested the complexity of relations between environment, disease, and citizenship in the development of the early twentieth-century US security state. This included the proliferation of the right to volunteer for experimental research as an element of medical citizenship. In public struggles over the proper terms of treatment and research, advocates for the interned were able to contest forms of apartheid that denied them control over treatments and reproduced their dependency on colonial institutions. At Moloka'i, this notably meant that patients at times publicly supported some head doctors who embraced experimentation, such as George Fitch, who hoped to lighten the strong arm of segregation and increase the availability of treatments.[71] The result was a governmental order that persistently articulated segregation as a tragic and necessary, yet hopefully temporary, measure that worked in the service of humanity writ large. This development is consistent with Foucault's point arguing that forms of sovereign exclusion can be augmented by the technologies of biopower, that surveillance and spatial mobilization of differences can reform the space of the camp. The combined power of separation, surveillance, and incorporation allowed nineteenth-century health authorities to "treat 'lepers' as plague victims, project the subtle segmentations of discipline onto the confused space of internment, combine it with the methods of analytical distribution proper to power, individualize the excluded, but use procedures of individualization to mark exclusion."[72] The Moloka'i settlements express such a technical conjunction because the government of species quickly shifted from one that conceived residents as prisoners in the 1860s to one that saw them, by the time of the US coup, as tragic patients who needed to be individuated, subjected to medical evaluation and experimentation, fed, clothed, policed, and provided with opportunities for spiritual redemption, entertainment, and contact—in short, subjected to a racialized colonial humanitarianism. By early in the first decade of the twentieth century, Progressive medical consensus was moving toward encouraging patients to identify and voluntarily confine themselves in leprosaria that would provide medical, social, and psychological benefits. This situation had an added dimension in Hawai'i, where the racialized enforcement of segregation meant that Moloka'i retained a thriving native Hawaiian community which was under attack elsewhere on the islands; Moloka'i could conjoin emergent forms of medical citizenship with the maintenance of a political culture that questioned the entrenchment of settler sovereignty. Patients were segregated, but their existence became the basis of identity in ways that tied them to their carceral space.

Given the racialized public representation of Hansen's disease in Hawai'i, it is no surprise that medical historians would view Moloka'i as a space of exclusion and that Agamben would take the case of vivisection as emblematic of empire's territorialization of bare life. The "leper island" has long been viewed sensationally as a space of pure abjection, and to figure those incarcerated as patients as being stripped to their biological condition is to perform the very act of rendering them zombies, "living dead," as Stoddard put it. Agamben's discussion does not account for the complexity of colonial architectures of control evident in the changing history of Moloka'i; even though the settler Board of Health successfully promoted vivisection during the late years of the Hawaiian monarchy, the annexation years witnessed the possibility that Moloka'i would intern all US American individuals diagnosed with the disease. In that example, state interests at times dovetailed with patient and community activisms that could assure skeptical publics of the humanitarian aims and modern techniques of the institution. There are thus reasons for attending to the distinct governmentalities of liberal humanism and fascism, even though Agamben is right to suggest that they exist within a single formation of power. Liberal humanism maintains a more subtle grip on the body in part through its ability to make coercion recede into the neutral domain of life itself and to align the subject with the government of species. The legacy of informed consent standard bears this out—patient populations have been progressively drawn into research in the intervening decades, arguing for access to experimental medicines as an aspect of medical citizenship rather than a simple abuse of power.

These details emerging from my analysis of the government of species in the context of Hawaiian annexation bear upon recent necropolitical theories that focus the camp as the privileged modern structure of living death. In *Homo Sacer* and *State of Exception*, Agamben identifies the camp—detention camp, concentration camp, medical camp, refugee camp, and so on—as the key spatial form of twentieth-century biopolitics. He argues, "*The camp is the space that is opened when the state of exception begins to become the rule.*"[73] In its more complex cartography of the modern/colonial distribution of death, Achille Mbembe's essay "Necropolitics" offers a breathtaking genealogy of sovereign "deathworlds" from the institution of plantation slavery to postmodern settler colonialism in Palestine. Elaborating the racialized architecture of the camp and dispensing with the simplistic generalization of the boomerang effect, Mbembe views "slavery" as foundational to the articulation of "biopolitical experimentation."[74]

Although these theories help us understand the persistence of sovereign power despite the apparent liberalism of some imperial states, necropolitical forces "are deeply linked to life-optimizing processes elsewhere in a biopolitical formation," requiring social analysis to "cross increasingly segmented biopolitical fields" of liberal and neoliberal government.[75] Jasbir Puar notes that despite its useful decentering of Europe, Mbembe's focus on the deathworld is "ironic" as it "excuses the investments biopolitics makes in mapping death in relation to the living."[76] Leerom Medevoi recalls Foucault's repeated reminders that political theory has focused too long on sovereignty to the exclusion of more subtle liberal-humanist operations of power;[77] one of the defining particularities of US empire has in fact been its confusing delimitation of sovereignty in a quest to secure power without taking responsibility for newly occupied territories or the peoples colonized there.[78] This delimitation of sovereignty was nothing new at the moment of Hawaiian annexation, by which time the differences between formal and informal colonial sovereignties had been used to dispossess American Indians of lands and treaty rights. In the context of transpacific financial globalization around the beginning of the twentieth century, the sovereign right to kill was one of several available methods of exercising state power. Any mapping of the coloniality of biopower must take account of the layerings of sovereign, disciplinary, and security mechanisms that work to contain the settler population against the multiplying risks of transborder and interspecies entanglement.

CHAPTER 2

Medicalized States of War

Venereal Disease and the Risks of

Occupation in Wartime Panamá

Only 200 patients with syphilis and 300 with other venereal diseases were more or less constantly under treatment. Among an indigent, mixed racial population, in which no coordinated program has ever been operative, it is reasonable to expect that perhaps 5%, or at least 15,000 persons in this particular population are infected with syphilis, to say nothing of the other venereal diseases. . . . The great reservoir of venereal disease in the population had not even been tapped.

—OLIVER CLARENCE WENGER, "The Venereal Disease Situation in the Panama Canal Zone and the New Control Program," 1943

Writing at the height of World War II, during a spike in venereal disease rates among US Army soldiers deployed in the Caribbean, physician and Public Health Service syphilis researcher O. C. Wenger was skeptical of official reports of low incidence of venereal diseases in the major cities of Panamá. Wenger is known today as the main government health official to assist the researchers who ran the infamous Tuskegee Syphilis Study (1932–72), in which six hundred African American men—mostly indigent sharecroppers—were used to study the progression of syphilis and denied proper medical information, informed consent, or the known effective treatments. This led to mass suffering and premature death among the study group. Having directed US Public Health Service campaigns against venereal diseases across both the southern United States (Alabama, Mississippi, Arkansas) and the army's Caribbean theater (Trinidad, Puerto Rico, Panamá), Wenger came to see race and class as primary factors in assessing the risk of sexually transmitted diseases. Explicitly condemning the Tuskegee group to an inevitable death, Wenger regularly demeaned the "darky" and "ignorant" black participants.[1] Thus it seemed apparent to Wenger, despite all available evidence presented by Panamá's health department and the army, that "an indigent, mixed racial population" living in the surroundings of the US-controlled Panamá Canal inevitably harbored dangerous bacterial species hidden beneath the skin.

If this was a conclusion that was geographically transferrable from Wenger's observations across the Jim Crow South and the US-occupied Caribbean, the language he used to describe the situation in occupied Panamá reflected the local history of disease eradication that permeated US public health efforts there. Wenger called the Panamanian population a "great reservoir" of disease that hid in plain sight, referencing the famed US campaigns against malaria that involved environmental warfare against bodies of water, which were literal breeding reservoirs for mosquitoes that transmitted disease. Invoking black and brown bodies as living reservoirs of risk for bacterial diseases such as syphilis, gonorrhea, and chancroid, Wenger subtly suggested that an eradication model of disease control enforcing forms of racial separation was necessary at the borders of the US-occupied Panamá Canal Zone. In this case, venereal diseases also required a gender separation—it was Panamanian woman in particular who posed the threat of border crossing, exposing servicemen made vulnerable by their innate sexual desires to the lure of "infected" women. If race indicated biological risk, gender added a social risk—the potential that any woman in public could be a prostitute who, mosquito-like, would target and exchange bodily fluids with the unsuspecting settler or soldier.

If in Hawai'i, US occupation revealed a missionary spirit for public health and tropical medicine as technologies that could be globalized in an era of expanding finance capital, the world wars presented different spatial challenges for disease in a global theater of US military deployment. This chapter explores how the spatial form of quarantine was transmuted into a broader technology of disease eradication in which US public health experts operating in the Caribbean worked to lock down small, sanitized spaces rather than to globalize disease control. This form for the government of species involved a type of environmental warfare that transformed interspecies relations through settlement, racial and gender segregation, and the control of tropical landscapes. It also involved what I call a medicalized state of war controlling how the soldier could effectively move and circulate across the borders of occupation while moderating the related bacterial risks of vice activities (drinking, sex) that seemed the inevitable consequence of such movement. While US public health in the Canal Zone and the adjacent cities thus involved a mix of extremely invasive techniques of territorial control and a mobilization of bodily contact and circulation through sex education, medicine, and controlled recreation, the last of these techniques proved the most resilient because it allowed public health officials to circumvent Panamanian and soldier resistance to the strong hand of public health policing. Nonetheless, the US military used all available resources to carry out an intense campaign to arrest and quarantine "venereal disease suspects," leading to the surveillance and criminalization of women who transited urban space. The campaign to suppress the sex trade in turn led to both private and public protests against women's arrest and incarceration in Panamá City's Venereal Disease Hospital. This history of occupation and conflict over the conduct of US soldiers in turn spurred a long-standing and conflicted association of US empire with vice, conflating the body of the sex worker with the US occupation itself.

Race, Eradication, and Occupation

Like Hawai'i in the era of annexation, Panamá during the thirty-year crisis of the depression and the world wars became a site for persistent concerns over disease. The US military attempted to control the key trade and military shipping route of the canal from other empire-driven states and to use it as an advertisement of US technoscientific progress. Where the settler colonization of Hawai'i had earlier been central to an emerging transpacific finance capital, the building of the Panamá Canal and the concomitant acceleration of transoceanic shipping made defense of the Central American isthmus a recurring concern for the US military, which feared penetration by Germany and Japan. In order to pave the way for canal building, white settlement, and military staging in the Canal Zone, tropical medical and sanitary interventions from Hawai'i, the Philippines, and Cuba had been transplanted beginning in 1904.[2] If, as Ricardo Salvatore has argued, the transoceanic waterway built from 1904 to 1914 was one of the key twentieth-century "transportation utopias" that articulated the spectacle of a new American modernity, public health at the Canal Zone provided a secondary technological spectacle, one displayed to millions of visitors of the Panama-Pacific exhibitions in California in 1914.[3] Health efforts in and around the US-occupied Canal Zone represented spatial eradication as a key model of disease control in the wake of the transcolonial war of 1898. Led by the invasive sanitary strategies of military tropical medicine, eradication in this case called for the environmental segregation of humans from microbes and from disease-transmitting species (such as mosquitoes) within particular spaces of settlement that the occupation authorities wanted to designate as "sanitated" against what they saw as the unruly tropical surroundings. As such, a variety of techniques of spatial segregation, chemical intervention, and education were central to maintaining boundaries for the management of movement and interactivity of bodies.

David Abernathy argues that the management of particular settler populations within a new territorially bounded sovereign zone was central to the notion of the success of the trumpeted efforts to eradicate malaria and yellow fever.[4] In the years following US settlement of the Canal Zone and the imposition of a system of racial apartheid, such notions of success were compromised given the inevitable failure to completely segregate settler populations from the objects of containment — groups of humans, insects, and microbes that were constructed as vectors of contagion. In contrast to Hawai'i, where Hansen's disease interventions were established within an increasingly global scope of missionary tropical medicine, this military-medical eradication strategy was focused on particular territorially bounded populations outside of which disease was left to proliferate widely. (Due to the racial profiling of disease risk, this logic persisted despite the fact that infection rates for many diseases were lower outside of the sanitary zones. Public health authorities often disavowed the fact that occupation itself was a cause of disease.) Instead of globalizing disease control on a missionary tropical-medical model, health reformers at the Canal Zone were concerned with regulating space, movement, and contact at smaller scales. As the United States entered two major global wars, imperial disease interventions reflected anxiety about territorial overstretch and resulting migration flows, making the small-scale and mobile containment of racialized populations and environments central to defining public health success.

The eradication model fit within a larger policy structuring space beginning during the canal building phase from 1904 to 1914. White settlers in the Zone adapted and expanded the terms of Jim Crow segregation imported from the US South by taking advantage of the occupation borders, a development that played a significant role in structuring Canal Zone–Panamanian relations during the wars.[5] As Stephen Frenkel claims, stereotypes of supposedly lazy, racially mixed Panamanians formed a backdrop against which Canal Zone authorities established their brand of racial segregation. In addition to separating Panamanians from the US-occupied Zone, US authorities set up sanitary corridors in the main population centers. (They also pressured Panamá to establish similar zones in major cities.) Within these corridors, they established largely segregated towns for white Zonians and for Afro- and Indo-Caribbean migrant laborers. These forms of spatial segregation were the basis of institutionalized class distinctions under the infamous Gold and Silver Roll system, which paid black laborers on a significantly lower scale. At the same time, there was an ongoing "belief that Panamanians were somehow unfit to work for the Canal Organization"—a belief which was the basis of significant anger among Panamanians toward the Zone and toward the growing populations of English-speaking migrant laborers who settled there.[6]

In addition to separating Zonians from multiracial Panamanian and Caribbean populations, health efforts were aimed at taming a Panamanian forest landscape that was seen as an unruly, teeming jungle. Although Panamanian and Caribbean migrant populations were most likely to face commonplace health threats like pneumonia and tuberculosis, the Health Department of the Canal Zone persistently targeted so-called tropical diseases and venereal diseases due to their high incidence among whites and due to persisting fears of contact.[7] Strategies for insect-borne diseases were adapted from Cuba, where military physician and sanitary officer William Gorgas became famous for using invasive sanitary techniques that literally remade ecological space, draining bodies of water and implementing fumigation, netting, and water sanitation to minimize human contact with mosquitoes. Under Gorgas's leadership, the Canal Commission secured official Panamanian recognition of US authority to regulate health in the major cities of Colón and Panamá.[8] John Lindsay-Poland argues that public health aimed at "altering the habitat of both human and insect, an adaptation of nonhuman ecologies to fit social objectives."[9] Such changes included clear-cutting vegetation and covering bodies of water with oil. These elements of landscaping sanitary zones were seen as pivotal to paving the way for white colonization. While working for the Isthmian Canal Commission, Gorgas claimed in 1909, "Our work in Panama will be looked upon as the earliest demonstration that the white man could flourish in the tropics and as the starting point for effective settlement of these regions by the Caucasian."[10]

Panamá was a special case of US imperial health administration given the unique juridical structure of the Canal Zone and its border cities, Panamá and Colón. The Canal Zone was a sphere of US administration under colonial treaty rights established at the moment of the US-sponsored secession of Panamá from Colombia in 1903, but it was repeatedly argued by US administrations that the Zone remained under Panamanian sovereignty, despite the establishment of colonial settlement, a US district court, and military occupation. From early in the canal-building process, US authorities took over health and sanitary operations at the Canal Zone and its adjacent cities from the Panamanian Republic, even as this required regular interventions outside the borders of the Zone. However, given regular territorial incursions into Panamá by the United States from the 1850s through the 1980s, this is no surprise.[11] With the confused legal status of people living in the Zone as well as the unpredictable subjection to US rule of Panamanians living outside it, citizenship was rendered ambiguous in order to enable flexible forms of US rule. As the executive secretary of the Isthmian Canal Commission remarked in 1906, "There is no such thing as 'citizenship' in the Canal Zone."[12]

Programs to control venereal disease were an outgrowth of institutions that operated to secure the health of race- and class-stratified Canal Zone populations, but they were also accepted as part of municipal and state reform efforts by the Panamanian client state. With its delimitation of sovereignty in Panamá, the United States maintained an official treaty-based claim to defense authority in the Panamá Canal Zone; however, US authorities also asserted important controls beyond the borders of the Canal Zone through political, market, public health, and military channels. This included strict controls of sexuality that were established during and immediately after World War II, when incipient shifts toward increasing political participation and public rights for Panamanian women collided with military paranoia about rising rates of syphilis and gonorrhea among whites. Women who were present in public in urban space came to be seen by US military health authorities as likely sex workers at the very moment in which Panamanian feminists built effective suffrage and equal pay campaigns and challenged the racialized citizenship restrictions of the Arias dictatorship in the early 1940s.[13] Thus even as the feminist Left developed a broader critique of authoritarianism, patriarchy, and suppression of minority rights, the ambivalent citizenship structure between Panamá and the Zone directly influenced US official fears of race and gender mixing, as well as the possibilities for women to contest their depiction as the reservoir of infection feared by Wenger and other public health experts.

Occupying Sexual Ecologies

Although malarial control and venereal disease control were primary concerns of civilian and military public health officials, emphasis on the latter increased substantially after the canal-building phase ended in 1914 and as growing deployments of US soldiers entered the Canal Zone. The spatial imaginary established through control of malarial mosquitoes deeply influenced cartographies of racial risk attributed to sex in Panamá, and a more general association of malaria and venereal disease was also common to wartime representations of transborder contagion.[14] At the borders of the Canal Zone, the jungle and the port city offered potent sites for the attribution of risk for sexually transmitted disease for US Americans who left the Canal Zone's sanitated corridors. By 1914, admissions to hospitals for syphilis and gonorrhea had surpassed those for malaria.

Soon after the canal opened, fears concerning the moral hygiene of soldiers deployed for World War I caused reformers to demand abstinence education programs. The first, failed attempt by the military to control soldier sexuality at the Canal Zone occurred in 1918, when the army quarantined soldiers, banning liquor and denying leave from the Canal Zone. In this effort, the moral panic was aimed at men's bodies, bodies understood as normatively uncontainable in sexuality and morally susceptible to incursions of outside vectors in an era of global war. Soldier rebellion quickly overturned the policy, leading the Panama Canal Department of the Army to seek policies more in line with US, British, and Spanish regulation (rather than suppression) of prostitution. While reformers in the 1910s outlawed prostitution in many cities in the United States, the regulation of sex work became the preferred policy for military base areas spanning Florida, Arizona, Texas, the Philippines, Haiti, Hawai'i, Puerto Rico, and the Canal Zone.[15] This was an outgrowth of what Laura Briggs calls the "international traffic in prostitution policy," which saw various imperial powers comparing and standardizing prostitution regulation policies in military-occupied zones across the globe.[16] In the Republic of Panamá, regulation of sex work consisted largely in the establishment of zoned red-light districts for legalized sex trades while sex work was carried out underground in other areas. Such policies were standard in major cities in 1904 immediately after the US-sponsored secession of Panamá from Colombia. Although health authorities at the Canal Zone sporadically sought to repress the sex trade in response to spikes in venereal disease rates, the general army policy was one of tolerance until the end of World War II, when the suppression campaign was carried out worldwide. For most of the first half of the twentieth century, tolerance of the sex trade was the preferred approach.

The large World War II deployment of US soldiers and civilians to the Panamá Canal Zone led to an unprecedented economic expansion in the region and an influx of migrant workers, including significant numbers of women from Europe, the Caribbean, and Central America. The influx of immigrants added to the already diverse ethnic and racial composition of Panamá, given the nineteenth- and early twentieth-century histories of US, European, Chinese, and pan-Caribbean migrations as well as a large presence of indigenous communities. During the war, the United States was able to convince the Panamanian government to allow the occupation of 134 new sites outside of the Canal Zone in order to defend the canal against Axis powers. Increased contact between the military and a transnational population of Panamanians, then, led to fears over miscegenation and vice. This was particularly true after the cessation of hostilities in 1946 met with a spike in venereal disease rates among US soldiers. Army health campaigns of the time stressed the international and multiracial character of sex workers in Panamá, in many cases denouncing the risks encountered by soldiers given the supposedly backward health and sexual practices of "mixed-race" or "Latin American" men and women. The United States lacked formal sovereignty but used treaty rights and aggressive diplomatic pressure to manage borders and control racialized populations to safeguard an idealized home front from the infections of occupation.

In 1942, the United States and Panamá established a cooperative venereal disease control program, initially set up as a system of registration and health surveillance for sex workers. This system openly tolerated soldiers' purchase of sex in Panamanian cities. In 1946, however, facing rapid increases in soldiers contracting bacterial infections of syphilis and gonorrhea, the military set forth an official policy to suppress prostitution in the armed forces worldwide. At this point, the cooperative control program shifted focus toward arresting and quarantining sex workers through the end of 1947, when the United States agreed to substantially reduce its military presence on the isthmus in the face of Panamanian nationalist criticism. Fears of bacterial venereal diseases drew upon US military and Panamanian anxieties concerning immigration and miscegenation, making possible repressive measures for the carceral quarantine of female bodies, new forms of regulation aimed at soldier sexuality, and transformations in Panamanian urban space. The targeting of Panamanian and Caribbean migrant women as infectious parasites suspected of harboring syphilis and gonorrhea furthermore impacted the broader relations between an imperial United States and emerging nationalisms on the isthmus. Facing increased migration and economic expansion, US health officials worked to institutionalize sanitary measures that presumed the contagiousness of women as reservoirs of disease.

Thus even as the official wartime deployment was ending, a new medical-police-military authority was established that criminalized women's bodies and thus constructed a particular, racialized figuration of "woman" that drew affective power from their feared interspecies contacts with sexually transmitted bacteria. The conjunction of new medical testing, sex education, modern pharmaceutical technologies, and the deployment of carceral quarantine worked in tandem to extend the state of war beyond the official end of hostilities. This medicalized state of war aimed strict regulations at women's bodies and women's public presence at the moment when a new regime of rights declared that they were full citizens of the Panamanian Republic. At the same time, new forms of soldier discipline were established that attempted to disrupt the smooth functioning of heterosexual desire during recreation leaves in the cities. Although male soldiers were seen as inherently and uncontrollably given over to sexual desires, these desires could be molded and channeled by forms of signaling that attempted to minimize soldiers' intimate contacts with the abyss of tropical microbes. The resulting unequal policing of sexuality—women's bodies were criminalized while soldier sexuality was tolerated and subtly disciplined—forged a unique formation of transborder and interspecies containment through the uneven distribution of fears and pleasures across the time and space of occupation.

Affective Prophylaxis

US military constructions of risk for venereal diseases involved sexual and racial surveillance as well as various forms of public signaling that I call affective prophylaxis. While in advanced stages syphilis manifests dramatic rashes covering the body, venereal diseases were often hidden beneath the clothing or limited, like gonorrhea and chancroid, largely to genital areas. Thus the medicalized state of war against venereal diseases had to do the work to visualize the hidden microbial hybridity of the venereal disease carrier and to guide soldiers in sexual self-regulation, especially during recreation leave. The technologies that disciplined soldiers' experiences of space and time made the prohibition of miscegenation an imperative of both soldier health and the preservation of national moral character. In the process, they incorporated normative knowledges of race and sex even as they at times had to minimize overt stereotypes of racial risk in order to guide the experience of urban recreation for soldiers.

From the perspective of military health, the figure of the Panamanian woman was made into a particular symbol of veiled contagion, a reservoir transmitting disease to soldiers and ultimately a biologized traitor to the Allied cause. Following World War II, the Panama Canal Department of the US Caribbean Command published a classified history of venereal disease among US soldiers that makes such associations clear. As justification for a variety of measures undertaken to control the sexuality of both US soldiers and women in Panamá during the half-century of US occupation, the report claims that Panamanian sex workers were "100% infected or infectious." The document, titled *Control of Venereal Disease and Prostitution in Panama*, names the bodies of Panamanian women as "reservoirs of infection."[17] Whereas Wenger used the trope of the malarial reservoir to refer to the Panamanian population at large, in this instance the specific body of woman was the microbial breeding ground, one that could infect and reinfect US forces over time. There was already a strong connection among tropical medical experts between the granulomas of leprosy and the chancres of syphilis; although the skin presentation and progression of the diseases were different, both of these bacterial infections connected fears of the tropical environment to sensationalized sites of oral and genital touch that treaded the borders of the erotic and the phobic. By adding the malarial trope of a female reservoir of bacteria, the discourse on syphilis in US-occupied Panamá aligned with fears of bodily penetration and exchange of fluids accomplished by the mosquito vector. Women themselves became living vectors requiring militarized intervention. Venereal disease was widely represented as a feminized traitor or Axis agent in army propaganda and public health campaigns.[18] A globally circulated poster in the late years of the war effort notes that a woman "may look clean—but pick-ups, good time girls, and prostitutes spread syphilis and gonorrhea." In the Canal Zone context, a Canal Commission circular from around 1920 parlayed the deathly touch of the leper discussed in chapter 1 into the figure of the Panamanian sex worker: "Warning! Beware of Whores! Nine out of ten of the prostitutes ashore are diseased. . . . Some of them look pretty good on the outside. They've got the female lure that puts a quiver down your backbone. But inside! . . . They're putrid, simply foul with disease. . . . You wouldn't rub up against a woman that was covered with leprosy. Well, many of these women have the leprosy inside." Contrasting the surface "lure" of the prostitute with the "leprosy" that made her "putrid" within, the circular figured the sex worker as bearer of an invisible taint of bodily mutation and degeneration. The sexed female body was both seductive and inhabited by unruly microbial life. Situating male soldier desire as preceding perception of sexual risk, sex education attempted to harness the categories of race and sex to intervene in sorting sexual objects, publicizing the gothic image of the leprous sore penetrating the inner life of the suspected prostitute.

To contain what it depicted as the exotic and seductive form of the Panamanian woman, the US military enforced restrictive measures to combat disease. The vice repression effort under the cooperative venereal disease control campaign undertaken by the Republic of Panamá, under US military supervision, in the cities of Colón and Panamá from 1946 to 1947, subjected women to arrest as "venereal disease suspects" (Panama Canal Department 85). Women testing positive for disease were sentenced to hospitals for enforced treatment, in practice confining them for up to six months per arrest. US soldiers were banned from visiting addresses classified as brothels, and the Panama Canal Department surgeon organized surveillance of the sex trade with Panamanian police and military police. Meanwhile, the military carried out sex education campaigns and expanded its long-standing use of prophylaxis stations in an attempt to force soldiers to self-regulate. While the arrests and carceral quarantine attempted to criminalize women's public presence, gendered quarantine emerged within a broader formation of affective prophylaxis that attempted both to spatially segregate soldiers from potential sex partners and to distribute a number of cues within soldier recreation time that would interrupt the normative unfolding of recreation into unregulated sexual encounters. These cues include the posting of venereal disease contraction scenarios on barracks bulletin boards, the use of pass sheets requiring soldiers to self-report sexual encounters during leaves, declaring addresses of suspected brothels off limits, and an expanded number of chemical prophylaxis stations for use following sexual contact.

As Allan Brandt notes, the larger military health context figured venereal disease as a threat to an idealized white family and, in particular, white femininity. Brandt discusses army health circulars that figure soldier venereal disease as a sign of failures in responsibility to both wives and mothers on the home front.[19] Roderick Ferguson has argued that one of the main effects of modern discourses of race is to universalize a binary gender system and normalize heterosexuality, and in these images an excessive and nonmonogamous soldier sexuality threatens the idyllic domesticity attached to whiteness (figure 2.1).[20] The racial schema of the cooperative venereal disease control program similarly linked sexuality and race by invoking the soldier's responsibility to white femininity and by figuring women of color as hypersexual or diseased, leaving an indelible racial taint upon the unfaithful soldier that would undermine his return to domestic space (home/land). Eradication thus

Figure 2.1 "For their sake . . . avoid venereal disease." Venereal Disease Circular, US Public Health Service, 1943.

exhibited a defensive logic of security—it would leave bacteria to proliferate among certain groups (racialized and feminized groups seen as inherently susceptible) but block the transmission of disease across lines of difference. Although it was the soldiers themselves who spread venereal diseases worldwide at the end of the war, sex education persistently figured women, who were always already potential prostitutes, as the site of risk.

Military discourse frankly figured the multiracial and international population of women present around the Canal Zone as sex workers who embodied gonorrhea and syphilis, seducing drunken and lonely US soldiers. The regulation of sexual labor thus became a key site for managing the public politics of gender.[21] In Panamá, US military officials tended to view the mestiza, Asian, and black women of Central America and the Caribbean as incurable due to racial and cultural predispositions. This was true from the earliest interest among Canal Zone officials in venereal disease. In 1918, the Health Department of the Canal Zone claimed, "Concubinage is as universal as it is among other Latin peoples. Illicit intercourse is not hidden with the mock modesty characteristic of the English-speaking races, but is frankly accepted as a necessity and is considered neither an evil nor a sin by anybody."[22] In the military report on the history of venereal disease among soldiers in Panamá, the author claims that although there is somewhat more concern about venereal disease among the "upper classes," "neither venereal disease nor prostitution is particularly feared by the Latin American population in general" (Panama Canal Department 13). The report goes further to blame Puerto Rican soldiers deployed at the Canal Zone as the cause of the particularly high rates of disease in the Zone. These soldiers are seen as exemplary of Caribbean moral failure: "The variable character of the Puerto Rican factor is connected not only with the number of Puerto Ricans, but with their type. The record indicates that the wartime Puerto Rican soldier's reaction to anti-venereal disease education was sluggish. Island troops are not easily impressed with the importance of avoiding infection" (104).

The report takes aim at transnational migrations that emerged at the moment of the expanded US presence in Panamá during the war. Although during World War II a new wave of Caribbean migrants entered Panamá to provide labor for a variety of occupations, the report focuses on what it sees as a covert influx of sex workers. The passages characterizing immigrant women as sex workers construct the sex trade as a racial threat to US soldiers and white Zonians:

Prostitutes encountered in Panama are highly international in origin. Although there are large numbers of Panamanian women ranging from complete African characteristics to light café con leche complexions, the Republic has always attracted women from abroad. Before the war many women were brought to Panama from Europe and there was considerable "white slave" traffic in the rotation of prostitutes from one Latin American country to another. To supply the demand of Americans, both civilian and military, white prostitutes have been very largely imported. During World War II travel restrictions curtailed the free passage of prostitutes from country to country, and new arrivals were from Cuba, Venezuela and other Latin American countries close enough to Panama to make human cargo smuggling possible. In an effective wartime step taken by Panama to reduce the incidence of venereal disease, a campaign was centered in rounding up and deporting all prostitutes who were found to be illegally in the Republic. It is safe to assume that imported prostitutes arrived in Panama suffering from venereal disease because, as indicated above, the original assumption is that prostitutes are 100% infected or infectious. (13)

The passage conceives of race as a key factor in determining the demand for sex workers. Viewing Panamanian women as representative of many of the races of the world, the author situates the US customer as the basis of the spike in demand for sex workers generally, as well as for white sex workers via other Latin American countries. While the rhetoric here may suggest the presumption of white soldiers' desire for white women (always already conceived as sex slaves), the report consistently takes note of soldiers' sexual relations with the full spectrum of racialized women, and mention of the "café con leche complexions" exoticizes mixed-race Panamanian woman. Meanwhile, the deportation of immigrant sex workers that the passage welcomes refers to a broader redefinition of citizenship along racial-nationalist lines after the Arnulfist constitution of 1941, which affirmed "Hispanic culture" and Spanish-Indian *mestizaje* to the exclusion of racial and ethnic minorities. The 1941 constitution would strip citizenship and immigration rights for Chinese, Arabs, South Asians, Caribbean Islanders, and Jews as "raíces prohibidas," allowing large numbers of deportations across immigrant groups.[23] Figuring immigration as a method of race mixing that increases the risk of venereal disease, the report displays anxiety about how the US presence may expand the sex trade and idealizes the Republic's efforts to circumscribe citizenship through its policing of race.

Military officials name miscegenation as a form of vice, directly linking disease risk to race. While US officials generally viewed "Latin peoples" as morally lascivious in contrast to white Anglo-Americans, the report also expresses a sense of disgust aimed at Afro-Latinas. Linking soldiers' cross-racial desire to the psychedelic self-expansion wrought by liquor during recreation time, the military historian describes "cantinas" in outlying areas of the Canal Zone housing "far from handsome" women of "either negroid or mestizo types whose charm would have been apparent to few but the well-alcoholized soldier" (46). Alcohol becomes vice in part because it weakens the defenses of the recreational soldier against a miscegenation understood as an aberrant, excessive, altered state of desire. This recalls Wenger's emphasis on racial mixing in the epigraph, where Panamá itself is a "great reservoir" in which race overdetermines venereal disease risk: "Among an indigent, mixed racial population . . . it is reasonable to expect that . . . 15,000 persons . . . are infected with syphilis, to say nothing of other venereal diseases."

Yet the presentation of race as a sexual risk was deployed within a broader logic of eradication that connected sanitary zones to racially segregated white space. This spatial logic would attempt more broadly to separate male soldiers from women of all racial designations. Viewing sex itself as an interspecies contact zone in which viral and bacterial transits occur in the interpenetration of male and female bodies, venereal infection becomes a moral incitement to contain sexuality. Despite the health officials' emphasis on race in the description of epidemic venereal disease, military sex education discouraged soldier contact with all women in Panamá, and tended to suggest that all Panamanian women were notably public women, transfer points in the transborder crucible of microbial outbreak. They were thus divorced from the privileged zone of the domestic Anglo-American couple idealized as the site for the heteronormative reproduction of the home/land.

As reminders of the hidden horrors of venereal disease, case histories were posted in barracks as evidence of the consequences of failed prophylaxis. Soldiers would see these posted histories on a daily basis as well as upon exiting military installations for leave time; the content of case histories was important, as was their existence as repeated visual cues articulating a warning against venereal disease as soldiers exited barracks (a space of regimental, collective time) and entered cities (for the increasingly policed affective release of individuated leave time). Military sex education officials presented histories that covered the broad spectrum of women's racial and ethnic classifications, detailing the risks of public encounter that military men of all ranks would inevitably face on leave:

> Pvt — had intercourse with a part-Chinese prostitute in a house near R—A—. He used all precautions. Unfortunately, he played around awhile before using condom. Result: Syphilis.

> Pfc — exposed himself to black prostitute in a native village while on pass. He used no condom or prophylaxis of any kind. He denied exposure on the pass sheet. Result: Gonorrhea.

> Lt — had exposure with "Blue Moon Girl" at P— T—. He had been exposing himself to this girl for five months and had never had any trouble. He did not think, under the circumstances, it was necessary to use a pro[phylactic]. This is not the type of reasoning that one expects from a soldier of his status. Result: Gonorrhea. (Panama Canal Department 60–61)

These examples stress the multiracial character of risk in military-civilian sex, even as they attempt to guide soldiers through a set of public encounters with Panamanian women. Such propaganda establishes direct messaging for soldiers, articulating exemplary discipline through the negative examples of failures to control the body during recreation time. This includes failed discipline that might result from the presumption that white sex workers are disease-free. Yet beyond these forms of direct messaging, the barracks posts also attempt to narratively contain the diversity of encounters a soldier might imagine while on leave, reducing the play and possibilities of sexuality to a number of expected types: visiting women in urban brothels, buying sex in villages, establishing repeated meetings with an off-again-on-again partner known as a Blue Moon Girl. Risk of venereal disease could emerge anyplace in soldiers' navigation of spaces inside and outside of military zones, in encounters with all sorts of bodies. Notably, these educational materials do not suggest sexual assault or homosexuality as possible forms of encounter. Military moralizing may have figured male soldier sexuality as excessive and inevitable, but could not conceive of a Panamanian woman who would deny a soldier's advances or else same-sex encounters either within or across racialized groups.

The reference to the Blue Moon Girl suggests a certain limit to what was imaginable as sex and intimacy from the military-medical perspective. The Blue Moon Girl or Good Time Girl was a category of woman defined by the army that included cabaret hostesses and other women who might, on occasion, have sex with and receive gifts or money from soldiers. This category allowed military sex education to contain the disruptive potential of dating or other transborder intimate relations that were not strictly prostitution within a logic of capitalist exchange. Military representations of Panamanian women tended to portray all women encountered in public as prostitutes, signaling the incomprehensibility of intimate attachments that could rise to the privileged status normatively attributed to the married couple, a couple presumably bonded by love rather than money. Amalia Cabezas, writing on academic scholarship on tourism that similarly emphasizes the sexual and monetary aspects of encounter, claims that such foreclosures fail to "interrogate our opposition between love and money." Noting that in "cross-cultural encounters" racial and class factors suffuse this opposition, Cabezas queries further, "What of situations that are not clearly marked as commercial endeavors? Or where sex is not present at all, but where gifts and other financial reward play a prominent role? What of relationships that combine pleasure, intimacy, and monetary support? The inability to read the subtle and liminal aspects of these encounters and an overemphasis on the sexual component risk privileging the sexual component as the most important aspect of interpersonal relations."[24]

If the brief messaging of barracks sex-education posts enacted one important kind of visual cue that would disrupt the unfolding of desire in recreational time-space, pharmaceutical technologies were also significant, enabling both a last-ditch effort at prophylaxis after sex and allowing for an expanded medical justification for carceral quarantine of women. Throughout the world war deployments, chemical prophylaxis was the standard treatment to respond to self-reported incidents in which sexual encounters placed soldiers at risk for venereal infection. Having set up prophylaxis stations in the cities and on bases, military officials required soldiers to visit them for localized injections of mercury (or other chemical preparations) within four hours of unprotected sexual contact.[25] While such chemical preparations were generally ineffective in the prevention of bacterial transmission, this invasive medical technology disrupted the smooth return of soldiers from recreation to regimental time. This meant compressing the remainder of leave and increasing the stress of return, potentially counteracting the psychedelic expansion of the self through drinking and other drug consumption. While this may have deterred some sexual encounters or encouraged increased use of condoms, the chemical prophylaxis requirement may also have provoked some soldiers to deny contact and evade the prophylaxis system altogether, as the military history report repeatedly notes. At the same time, prophylactic discipline made sex an end to recreation, which might actually increase its excitement or intensity as a final whim of leave. These connections between drinking, sex, soldier leave, and sex education—which simultaneously link immune, sexual, digestive, and nervous bodily interfaces crossing the borders of the Canal Zone—become the targets of forms of discipline that attempted to marshal soldier desires toward ends that would hypothetically contain venereal disease. Such policing demonstrated a much more developed form of containment than the short-lived attempt to quarantine soldiers to bases in the Zone during World War I.

Sex education, pharmaceutical intervention, and carceral quarantine played key roles in the Panamanian articulation of US empire's government of species. Race had a powerful justificatory presence in the military logic for aiming carceral quarantine at women, even if it became important for military sex education to diminish race as an epidemiological risk factor in the work of promoting disciplined relations to soldier sexuality. Race became a marker of the diversity and dispersal of risk as well as the complexity of encounters in which extramarital soldier heterosexuality could be expressed and where the body could become vulnerable to unseen interspecies intimacies. It was thus closely tied to the imagined scenarios of growth in the sex trade, as health officials attempted to describe what they saw as economic migration factors tied to aberrant hygienic and sexual practices among Panamanian women specifically, and Latin American and Caribbean populations more generally. Thus disciplining soldier sexuality—for an increasingly multiracial army—overlaid the Jim Crow racial structure of Zonian-Panamanian economies of border tourism. Racial vision—soldiers' affective sorting of the diverse possibilities of sexual object choice—needed to be exercised but in turn contained by spatiotemporal techniques that regularized the prophylactic ritual of cleansing and that incorporated the multiraciality of Panamanian women and soldiers into recreation time's controlled expenditure of affective release. The venereal disease quarantine—the space of the carceral camp—was but one spatial articulation of an increasingly complex flow of police, environmental, and health powers used to restructure transborder and interspecies relations.

Transit, Transition, Incarceration: Immobilizing Woman

Feminist scholarship on sex work is often divided between researchers who frame sexual labor in terms of multiple social forms of oppression and those who insist that sexuality—including sexual labor—must be viewed as a site of women's agency. Svati Shah historicizes this opposition going back to late nineteenth-century moral panics that sensationalized the "white slave" as violated and trafficked woman.[26] Even as this opposition frames contemporary debates over sex trafficking and sex work[27]—which often reproduce a false image of power that opposes victimization and agency—I wish to document the ways in which the suppression and regulation of the sex trade in Panamá worked to respatialize urban life and create sites of political emergence in which the figure of woman was ambivalently aligned with nationalist projects critiquing the US occupation. As new efforts to discipline the space-time of soldier recreation both normalized sex work and circumvented the suppression of sexuality, the carceral quarantine of suspected prostitutes transformed the ways in which sex spatialized the city and generated anticolonial nationalism.

As the expanded US wartime occupation increased the presence of soldiers in the cities and brought about a new wave of immigration, depictions of Panamanian cities emphasized increased crowds, dollars, and English speakers. In this context, the cooperative venereal disease control program brought increased surveillance of the expanding nightclub scene and formal sex trade, decentralizing a segment of the trade through an expanded network of taxi drivers and illegal brothels. At the same time, dragnets carried out against "public women" at night meant that sex work was increasingly carried out in the daytime and in areas outside of the red-light districts, while women having no relation to the sex trade were forced to circumvent surveillance of public spaces at night. As a whole, these transformations impacted the forms of life in the city as well as the forms of sex tourism. Even as the increased transborder contacts transformed urban space, they also diversified the timing of sex based on the relations between soldier leave time, nighttime curfews, and relocation of sex work into cabs and into distant or underground brothels.

Newspaper coverage in 1946 and 1947, the years during which the US-Panamanian cooperative venereal disease control program carried out its prostitution repression policy, only made sporadic reports of antivice activities. These reports often centered on pressure US officials placed on Panamanian health authorities, impacts of vice repression on local businesses, or the relatively high incidence of venereal disease among white Canal Zone and military populations. Articles most often appeared in the Spanish editions of papers, with the English versions for Canal Zone audiences generally disavowing US soldiers' roles in the spread of venereal disease and the invasive nature of quarantine. In the May 1946–December 1947 issues of *El Panamá América* and *La Estrella de Panamá*, a number of articles highlighted national and regional antivenereal efforts, government policy, perspectives of elites and business owners on the restriction of soldiers, and, at times, nationalist perspectives condemning stereotypes regarding Panamanian disease. *The Panama Tribune*, a newspaper aimed at the English-speaking Afro-Caribbean readership, carried many stories critical of US sanitation efforts. It also ran two stories at the end of 1947 that emphasized the sexual immorality of white Zonians, countering Wenger's Jim Crow–influenced colonial discourse that attributed venereal disease risk to race. The paper claimed, "United States servicemen are the main supporters of these brothels and consequently the greatest sufferers of venereal diseases."[28] A number of news reports in the first days of May 1946 reported on the military's establishment of "off limits zones," new immigration restrictions, and undefined forthcoming measures representing "drastic means" for the repression of prostitution.[29] These articles reflect skepticism about public health recommendations in the antivice campaigns.

While chemical prophylaxis allowed military medicine to intervene in the lives of soldiers in defense against venereal diseases, pharmaceuticals also underwrote the deployment of quarantine authority against Panamanian women. The use of penicillin was standardized by the military as of 1943 (and globally by 1948). Venereal disease control advocates then had the first reliable method for combating bacterial infections, and along with it, justification to mandate enforced quarantine. Penicillin was administered on an inpatient basis by IV, meaning that health officials had medical justification for maintaining women in carceral quarantine for the duration of treatment. The cost of a dose of penicillin fell rapidly in time for deployment at the end of the war.[30] Penicillin was immediately made available to servicemen in the Canal Zone in 1943 and was distributed to incarcerated women in the 1946–47 program (Panama Canal Department 77, 80). However, penicillin also caused a panic over the moral status of the soldier, who would not receive nature's supposed reprimand for sexual deviance in the form of an alien bacterium.

These changes differentially impacted soldiers and Panamanian women, who bore the brunt of the suppression of prostitution policy through curfews and enhanced surveillance as well as the lost time of arrest and the six-month maximum duration of quarantine. The 1947 military report details how women themselves circumvented arrest and challenged incarceration in the face of such regulation. In the report, one official acknowledges the legitimacy of local protests against the cooperative control program based on civil rights concerns. Explaining that military police organized patrols across Panamá City, the official claims, "Women and girls found on the streets late at night were arrested in droves and taken to Santo Tomás Hospital for examination. If found to be infected with venereal diseases, they were sentenced to the Women's Hospital for cure. Females who had legitimate business and were enroute home from night jobs or after visiting friends were frequently picked up in the drag net, and complained loudly. They had a right to object" (Panama Canal Department 85). To the extent that the army negatively associated Panamanian women with racial impurity, promiscuity, low class status, and low status of sex education, it demonstrated a degree of sympathy in such passages for wrongly arrested women. It did so within the context of its advocacy for a tightening of gender roles, thus idealizing wrongly arrested women against those deemed guilty because they suffered from venereal disease. Disease itself, rather than evidence of a criminal act, became the basis for incarceration.

The limited news reporting on the dragnets operates according to a similar logic. In a vague article appearing in *El Panamá América* on June 28, 1947, the paper claims that regular "nighttime pickups" beginning in the early evening in Colón "cause panic in commerce and unrest among people of good standing." Focusing initially on business protests against "the constant persecution" at cantinas, clubs, and theaters, the article goes on to state that the authorities are attempting "to reprimand immorality without causing harm to innocent people." Shifting tone, the article then moves from the fear spreading among the inhabitants of Colón, who increasingly stayed home after dusk to avoid arrest, to paying compliments to the police's discrimination in enforcement. Without ever directly mentioning sex, prostitution, disease, or US military pressure, the anonymous author skirts an open attack on the police, city and state officials, and business owners. Instead, the article progresses through a binary logic pitting "immorality" against "innocent families" scared away from the city's commerce.[31] Similarly, when the author of the military report claims that "the promiscuous women enjoyed the experience no more than their innocent sisters," he identifies them as "crafty operators" who knowingly circumvent the law: "To avoid possible sentence to the venereal disease hospital for treatment, promiscuous women became crafty operators and contrived to escape detection by an informal and widespread recourse to the taxicabs of the various cities and towns in Panama" (Panama Canal Department 85).

Reading the military report against the grain of its moralizing, a variety of responses emerge to the management of urban space and time under the venereal disease quarantine and suppression effort. Once in the detention facilities at the Women's Venereal Disease Hospital at Hospital Santo Tomás, the report casually notes, "the tempestuous Latin American prostitute acts first and thinks later." Rebelling against attendants and occasionally fighting other inmates, the women only received rounded utensils, because "patients are inclined to sharpen knives and forks" into weapons (12). Women involved in the sex trade were increasingly forced underground. Those arrested protested against arrest, examination, and confinement, and women quarantined in venereal disease units refused to follow rules and to properly follow treatment schedules. As such, the report claims that antibiotic treatments worked at a lower rate for incarcerated women at the hospital quarantine in Matías Hernández (80).

Privileging such moments of interruption to the smooth functioning of urban surveillance and carceral quarantine throws into relief the more widely publicized Panamanian state and business sector protests of the program that, though contesting US control over Panamanian life, were structured by both a bourgeois moralism and a strictly gendered notion of the public. *The Panama American / El Panamá América* documents a number of government and business protests of the cooperative control program. In 1946 and 1947, the paper ran articles on the impact of the military policy on commerce. In protest of the military policy, bar owners in Panamá City planned to strike for one week demanding the waiver of licensing fees.[32] Meanwhile, as tension between the US and Panamá heated up over the continuing US presence at military bases outside of the Canal Zone, and as Panamanian officials began to tire of the continuing pressure over the sex trade by US officials, Spanish-edition newspapers reported that Panamá was doing better at combating venereal disease than the United States and US-controlled Puerto Rico. They cited Dr. Guillermó Garcia de Paredes, director of the Department of Public Health, who claimed that a strong public health program lowered venereal disease rates to "a level far lower than that of Puerto Rico and many cities of the United States"; furthermore, US efforts to suppress the sex trade had only "stimulated clandestine prostitution."[33] Paredes made a more detailed case against repression of the sex trade at the Twelfth Panamerican Sanitary Conference in 1947. Again emphasizing that "during the most critical period of the European war, when with the cooperation of the military authorities of the Canal Zone, we intensified our campaign of periodic examinations and segregation of ill prostitutes," infection rates fell below those in US cities.[34] Paredes makes sure to point out the particular situation of the canal region, emphasizing "sites of large population and migration, in which exist a relative abundance of single men, or those separated temporarily or permanently from their legitimate spouses" (3), to make a case that the US occupation was the central cause of the proliferation of venereal disease. Yet he also constructs female sex workers as suffering from "ignorance or mental deficiencies" (4).

News reports did, occasionally, glimpse women's public refusal of the carceral policy. In an article published on May 23, 1947, in *El Panamá América*, the newspaper reports a roundup of "public women" across the city of Colón.[35] The article decries the immorality of three aspects of the sex trade. The first is the presence of school-aged girls at the cantinas. The second is the identification of the "illness" of twelve of the arrested women, who were subsequently detained as the rest of the women were set free. Finally, the article reports on what it identifies as a "disagreeable spectacle": at 8 AM, as the women were brought to the clinic for testing, a large group of onlookers had assembled, including a group of schoolchildren, who witnessed the escape of one of the women. The crowd yelled, "Catch her!" Amid the excitement of the crowd, the police detained the woman and forced her to enter the clinic. The spectacle served to sensationalize the sex trade as the problem was narrated as a question of proper policing. Although the unsigned article presents an antivice perspective, it raises the issue of escape attempts. The military historian who wrote the 1947 report does not provide statistics on escape attempts, but does parenthetically mention that "escape" was the only way of avoiding hospital sentences for women who tested positive (Panama Canal Department 12). Such discourse conceptualized venereal diseases within a logic that made women's bodies the primary transit points of venereal disease.

The Fallen City: Vice Discourse, Contagion, and Anti-imperial Literature

Whereas newspaper reporting on the cooperative venereal disease control program failed to capture the complexity of interests, technologies, and bodies involved in the military program, literary representations of Panamanian cities have a long history of grappling with the imperial politics of sex. In this final section, I explore how the affective transformations of urban sexual life under occupation contribute to the literary registers of anti-imperialist nationalisms. The military representation of Panamanian women was deeply interlinked with moralizing antivice discourse that impacted authors who engaged with the anti-imperialist traditions of midcentury Latin American modernism. Writers often figure Panamá as being wholly dependent on US American dollars, feeding a national culture of corruption. Such depictions, which I term the "fallen city" trope, resonate with Frantz Fanon's reflections on neocolonialism, which state that the managerial class that controls the national leadership will "set up its country as the brothel" or "bordello" of the external power.[36] The response is often to portray, as does Guadeloupean novelist Maryse Condé, either Colón or Panamá as a fallen, dystopian city run on US dollars and sexual vice.[37] Such representations do work against the US military's depiction of racially mixed Caribbean and Latino/a populations as inherently lascivious and diseased, and they attempt to center the ways in which settler-colonial projects targeted women's bodies and labor. At the same time, they restage a colonizer/victim paradigm, drawing affective power from the opposition between women as agents (of prostitution) and victims (of trafficking or economic inequality). This move obscures the particularity of Panamanian women's responses to military occupation (including their escape attempts from quarantine and their circumvention of police surveillance) and the ways in which anticolonial nationalism assimilates normative knowledges of sex, gender, race, and species. Given that, as Alan McPherson has noted, the figures of the mistress and the prostitute came to represent imperial injury to Panamanian identity, nationalist anti-imperialism was highly gendered in ways that sought to masculinize and desexualize resistance to US power. Visions of urban violence and monstrous interracial sexualities mark literary accounts of both the contagions of urban life and the corruption of occupation. Equations of sexual labor with imperial vice repeat the logics of Progressive moralizing against the sex trade and shore up anticolonial nationalism's heteropatriarchal gender roles.[38]

Although women are seldom foregrounded in Panamá Canal literature,[39] sex workers repeatedly figure in the plots of a handful of relevant texts depicting the occupation. A text explicitly concerned with the effects of migration and sexuality on women in Panamá, Maryse Condé's 1987 novel *Tree of Life* begins with the story of Albert and Liza, the narrator's grandparents. Albert, a sugarcane cutter who left Guadeloupe for Colón to become a canal digger, was recruited by an agent at a brothel. He spends all of his earnings in the Colón red-light district after his move to Panamá. But Albert eventually leaves the brothels to marry Liza, only to experience the violence of racially segregated Canal Zone hospitals when Liza is unable to obtain proper medical treatment and dies in childbirth. As such, the brothel and the hospital represent the false promise of US capital.

While Condé's historical novel is relatively brief in its treatment of the sex trade, earlier Central American modernist literature written during the thirty-year crisis invoked the dystopian vision of urban sexuality in greater detail. Written during the expanded wartime US occupation, Rogelio Sinán's

1947 Spanish-language novel *Plenilunio* (Full Moon) more directly associates the US presence with vice.[40] As Sinán's first novel, *Plenilunio* focuses on the cultural memory of the dislocations produced by US expansion in Panamá. While it is possible to read Sinán's novel as an expression of literary nationalism, *Plenilunio* also gives an account of how alienation and violence undergird modern political projects and become particularly visible at sites of transborder contact such as Panamá generally and the canal in particular.[41]

Living in Panamá City, the novel's characters chronicle the sordid details of their lives in a city funded by American dollars at the behest of US soldiers. The elderly Don Céfaro, like Condé's Albert, becomes hardened after losing his wife in childbirth.[42] Turning to a life of pimping and drug trafficking, he recounts the pernicious US influences during World War II: "The coffers of Uncle Sam poured out for the cause of war. . . . A few drops fell on us in the Isthmus—a multitude of drops that the many greedy and thirsty gathered. I was blinded. I had no scruples about wasting my money on prostitutes. They had arrived at the Isthmus from all over! Mexicans, Cubans, Argentinians . . . they arrived from all over!" (40).[43] The son of a cabaret dancer and a US soldier, Don Céfaro's narrative of his involvement in the sex trade, drug trafficking, and other crimes places sexuality at the center of the violences that occurred on individual, familial, and national levels when the canal opened to US influence.

The protagonist of the novel, Elena Cunha, is the mixed-race daughter of a cabaret dancer, who speculates that her father may have been a US soldier (36). She expands on the critique of the moral decline of Panamá and the Americas under US domination. For Elena in Panamá City, "Everything was drowned: Morality, in skirt, singing hymns inside brothels. . . . There was an exaltation of egoism, of sex, of prostitution" (69).[44] Having personally witnessed the violence of prostitution, pimping, and drug trafficking, Elena is described as mentally ill, and she invokes the spiritual fall of the Americas under US empire when she describes "bad angels flying over America" (72). Sinán's novel thus situates its critique of US excesses in terms of the moral and psychic impacts of empire on Panamá. Like Joaquin Beleño's influential novel *Luna Verde* (1950), Sinán conceives of the fall of Panamanian sociality in terms of a world of crime, greed, and corruption. Overall, the representation of the sex trade is conceived in terms of increased discrimination, lost morality, and capitalist greed under US imperialism—all of which are connected to the increased density of bodies occupying the cityscape and the related covert and public forms of greed and sexuality.

Ecuadorian writer Demetrio Aguilera-Malta's 1935 novel *Canal Zone* precedes Sinán on the gendered and racial logics of the fallen city trope. Aguilera-Malta was one of several writers from Guayaquil who pioneered social realist writing in the 1930s. *Canal Zone* is known as a novelistic critique of imperialism that, in particular, highlights the black diasporic presence in Central America. The character Pedro Coorsi, like Sinán's Elena, is the offspring of immigrants who arrive at the canal for the economic opportunities it offers. In this case, the parents are a drunken, abusive Greek sailor and his Afro-Caribbean wife. Aguilera-Malta feminizes the canal, figured as a mother to the nation: "a gigantic teat with thousands of nipples to calm the hunger of the masses."[45] This is initially embraced as progress, as the narrator witnesses the "transformation of a new Panamá," "a hygienic, fraternal, inviting Panamá" (12). In Pedro's childhood, Panamá is figured along the triumphant lines of imperialist representation as "the land where two oceans are united" (16). According to the narrator, however, the nights saw the corruption of the new Panamá in the sensational images of exposed flesh and alimentary disgust: "The cantinas and cabarets opened their eyes magnetically. They drew out their carnivals of flesh. They toasted their song of bottles. They drank in front of cemeteries. When the sun swept away the dawn, they vomited scraps of human meat" (14).[46]

After the Great Depression, mass unemployment shakes Pedro's vision of a prosperous Panamá uniting the world as greed and vice sweep the Isthmus. Over 40,000 US soldiers arrive in order to defend the canal; in the process, Panamá City is transformed by a constant song of two words spoken in English: "*Welcome, sailors!*" (116). The cabarets of Santa Ana district are the mainstays of business serving the marines, and "though the women were surrendered to trading their bodies by personal initiative, there was loot for the landlord, the milkman, the tailor, the journalist, the police, everyone" (119–20). In an economy that comes to rely on vice, even Pedro, who works tirelessly for black rights and supports a tenant's strike, sees his new job as taxi driver employed by a brothel as a sign of the total corruption of the isthmus. This corruption reaches its height when the narrator views the show *Exhibition*, in which nude female contortionists delight the marine audiences with "whatever the abnormal person's imagination could conceive" (120). In "deplorable English," the performers solicit "three ways" and engage in a "monstrous suck [*mamada monstruosa*]" (121). If oral sex, nonmonogamy, exhibitionism, and group sex are not signs enough of the evil of occupation, then Pedro's ultimate death is, as his passing is virtually ignored and the trade of the brothel goes on uninterrupted. Aguilera-Malta connects this ultimate death to the fall of a city expressed in the oral excesses of binge drinking, vomited meat, and blow jobs, all signaling the transition away from the apparent innocence of the narrator's youth.

Sexuality in such a contact zone, of course, carries risks of violence—the risks that pit marine against woman and that display the costs of migration. Each of the works I have discussed foregrounds the destructive aspects of US empire by representing the economic exploitation of Panamanians and Caribbean migrants, situating the sex trade as a negative impact of empire, and associating the fall under occupation as either a tragic victimization or a spiritual-moral failure of choosing money over independence. Condé's, Aguilera-Malta's, and Sinán's novels present the blurry and multiple stories of wartime urban life from the perspective of narrators reflecting on Caribbean dependencies on the United States, and, at times, experiencing the dissolution brought by the speeding of time and the transit of bodies under urban capital expansion.

However, other authors do present alternative versions of the fallen city trope. Eric Walrond wrote a large volume of stories concerning Caribbean working-class life.[47] One of Walrond's stories, "Godless City," does represent Colón as a space corrupted by greed and sex work, doomed to the biblical fate of repeated fires. Walrond represents the periodic fires that ravaged the city multiple times in the 1890s and 1900s in apocalyptic racial terms, depicting the colonial "whitening" of the city and the "plagues" that devastate it. The US naval men who speak in the opening passages describe it as such: "Every ten or twelve years the city's got to be cleaned—wiped out—destroyed. A plague— a fire—something. God's work. . . . Just acres of smoking ashes—ashes of flesh—ashes of bone—ashes of wood. White—all white! Z'got to be clean."[48] Initiating the story with an association of the city with immorality and vice, the narrator goes on to describe the "humming hell" that constructed US American accounts of the city and its infamous *barrio rojo*. There were "one-armed men," "knuckle-dusters," and the "living cancer": "Chinamen with leprous ulcers on their skins." Yet the licentious atmosphere of the red-light district was a result of "morals from higher up" (166)—the Spanish girls who loitered at the saloons were no different than the cabaret dancers from "France, Sweden, Germany, Cuba, Costa Rica, West India" who "danced and sang before the greedy eyes of the applauding *conquistadores*" in the affluent parts of town frequented by US servicemen (165).

Despite this invocation of the fallen city, Walrond here and elsewhere alternatively portrays an idealized subaltern community of migrant laborers that counters the alienating and oppressive nature of occupation. Celebrating the nights of dancing, the practices of obeah, and shops selling calaloo and fungee, the narrator claims that all of these scenes are obliterated in the fire that devastates everyday life in the city; there is a rumor that the fire started at the Red Raven, the cabaret frequented by white sailors (172). Situating the basis of the moral failure in the US occupation, the ironically invoked immorality of the migrant populations of Colón works to counter elite understandings of black and Asian bodies as sources of filth, pestilence, and sin. Similarly, a 1925 Walrond story, "The Voodoo's Revenge," represents brothels as a site of covert resistance to US-supported rulers. The main character, an editor of a progovernment newspaper of Colón, is a former labor migrant who worked his way into the media after escaping the canal. After being jailed by a judge of the ruling party, he turns to the powers of obeah, to carry out the poisoning of the judge; ultimately their connection at a brothel links the sex trade to the routes of Caribbean migrant cultures. In this case, the brothel is a site of male resistance to empire. In Walrond's 1940 piece, "Morning in Colón," women associated with the sex trade become part of Walrond's picture of empire. This very short story seeks to humanize sex workers and other laborers in Panamá by normalizing their sexuality and uniting them in shared forms of economic hardship in ways that do not sensationalize sex. The story is a dialogue between Rufus, a black creole ice salesman, and Rosie, described as "Chinese half-caste."[49] As Rosie approaches, Rufus teases the woman over her clothing, inviting Rosie's protest over treatment she associates with black women. But after these initial verbal jabs, the conversation quickly turns to the lack of supplies due to the US military presence, demonstrating that Rufus and Rosie are similarly disadvantaged by US imperialism despite their apparent class and racial differences.

In all of these works, relating psychic violence to social and ecological violence within American cartographies of power becomes a central purpose of the deployment of modernist themes and forms.[50] In this sense, the fallen city trope fits into a modernist logic that embeds discussions of violence across lines of social difference within a particular urban ecology in which bodies are rendered economically, politically, and biologically vulnerable by empire. This strategy at once opens new possibilities for the critique of US imperialism and forecloses certain possibilities for an appraisal of women's political emergence in a situation in which nationalism is normatively masculinized, and thus highly amenable to the moral logics of antivice policing.

The end of World War II would help bring an end to the US military's repression of prostitution regime, as Panamá pushed for total sovereignty outside the Canal Zone. With postwar economic turmoil, the figures of the mistress and the prostitute came to represent the injury to masculinist discourses of Panamanian identity. Since nationalist rhetoric from the 1940s to the 1960s associated the US presence with emasculation, nationalism was conceived largely in terms of economic and political independence — in terms of public rather than the often private, intimate acts that constituted the urban geography of commercial sex. Yet if we alternatively understand resistance to imperialism in terms of the government of species in wartime and postwar Panamá, we can begin to see women's public presence and intimate labors as sites of struggle in the spatial politics of empire and nationalism. Such struggles contested the logics of race and species that would define the behaviors and the destiny of the largely migrant population of Panamanian women in terms of a stereotyped, pestilent sexuality and innate vulnerability to infection.

In this chapter, I have argued that the control of venereal disease during the world wars demonstrated the consolidation of spatial approaches to disease intervention, moving away from missionary global approaches and suturing sanitation to local border controls based on detailed space-time architectures of racial apartheid and military occupation. Drawing on associations of female bodies with insect reservoirs and with a treasonous ability to seduce, this spatial emphasis required the development of forms of signaling in order to channel the self-expanding pursuits of soldier recreation, such as drinking and sexual tourism, away from interspecies contact. This in turn intensified soldier sexual encounters by interrupting recreation time, even as it restructured urban space in ways that criminalized women's public presence and subjected them to invasive screening and quarantine. The government of species thus involved a form of warfare that extended beyond the official end of World War II and managed pharmaceutical and police intervention into everyday life. In the process, military authorities criminalized women's public citizenship during the rise of new feminist citizenship struggles and exacerbated divisions with the Panamanian client state, opening highly gendered nationalist discourses that portrayed the US occupation as itself a kind of vice or moral failure. At midcentury, spatial approaches to disease control were encountering a racially and biologically complex world, which required establishing new forms of biopolitical segmentation in order to optimize the relations of populations and species transiting across the continental borders.

CHAPTER 3

Domesticating Immunity

The Polio Scare, Cold War Mobility,

and the Vivisected Primate

In January 1939, *Life* magazine published a striking image of a monkey (figure 3.1). The image centers the intense gaze of a rhesus macaque who, fur soaked, is half-submerged in coastal waters. With a heavily furrowed brow, the lone monkey casts a dark shadow over the rippling foreground, curtained by ominous clouds and a distant Caribbean coast behind. The editors gave the image the title, "A misogynist monkey seeks solitude in the Caribbean off Puerto Rico." The accompanying caption explains that the apparently distressed "bachelor" rhesus retreated to the water to escape "the chatter of innumerable female monkeys."

Life had commissioned Hansel Mieth—a photographer famous for her social-realist portraits of Depression-era life in the United States—to document a new primate research colony established on Cayo Santiago, a small islet in the southeast of the Puerto Rican archipelago. An earlier *Life* article on the colony had characterized the monkeys there as "domineering" because humans in their native India considered them "sacred," consolidating orientalist stereotypes that viewed Hindus as, on the one hand, superstitious in their views of animals and, on the other, unfit for self-rule.[1] Having been domesticated into the contexts of US settlement in the Americas, however, the editors' caption transfers this assertion of India's improper distribution of sovereignty between state, human, and monkey into a patriarchal identification with the animal against the supposed feminization of modern social life. The editors imagine the primal masculinity of the rhesus smothered by the "chatter of innumerable female monkeys," which signals the frivolous and sedentary excess of modernity. This was a common refrain of US idealizations of a muscular masculinity in the 1930s, when technology, urbanization, and

Figure 3.1 *A Misogynist Monkey Seeks Solitude in the Caribbean off Puerto Rico*, 1939. Photograph by Hansel Mieth. The LIFE Picture Collection / Getty Images.

female labor participation were forces commonly attributed to the degeneration of white settler life after the closing of the continental frontier. In the logic of the editors' joke, this monkey's penetrating gaze signals a masculine depth of feeling contrasted with the overexcited surface affect of the feminine.

Mieth rejected the editors' sexist response to the image. However, her own description of the rhesus as a refugee stranded on an island in a time of war invokes a similar vision of primate subjectivity.[2] The monkey's depth of emotional character is made visible in the comportment of the body and its engagement of the photographic gaze. Both Mieth's and the editors' interpretations play on an idea of geographic and social estrangement experienced by the exiled monkey. The reading of the monkey as angry, dissatisfied, or impatient relies on the engagement of the monkey gaze by the camera lens, the animal's active participation in an interspecies visual economy. As Megan Glick explains, the ability of photographic representation to domesticate the non-human primate into the sphere of whiteness depended on the attribution of capacity for sight, with a focus on three particular characteristics that were increasingly associated with photographed chimpanzees beginning in the 1920s: literal capacity for sight expressed through eye position and distance perception; the ability to meet the human gaze; and attraction to the photographic process itself.[3] This training of the monkey's gaze upon Mieth's lens allows for a portrait of an individual, feeling monkey, rather than simply a documentary photograph of an animal specimen. This monkey moves, in both the physical and emotional senses of the word, swimming unexpectedly away from the site of incarceration, angrily uniting emotion with motion.[4] To the extent that the rhesus in Mieth's photograph appears to engage directly with the photographic gaze and to demonstrate an interspecies economy of visual and affective exchange, it allows the Indian-origin animal migrant to be domesticated into visions of wartime social and geopolitical alienation in the Americas.

Mieth's striking image thus captures in the primate body a conjunction of mobility and feeling, two privileged capacities of the nervous system that were increasingly significant in an era of scientific medicine.[5] The particular monkey Mieth photographed was one of 409 imported from India to Puerto Rico to found the first free-ranging research colony of rhesus macaques in US-occupied territories. In 1939, the rhesus was fast becoming medical science's key experimental model of the human, partly through its central use in research on poliomyelitis. The use of rhesus in polio research led to a scramble for laboratory monkeys among US and European researchers and pharmaceutical producers, forming the beginnings of a primate trade connecting US research laboratories to the Caribbean, Asia, and Africa. Yet even as rhesus were increasingly domesticated as feeling, almost-human subjects ripe for laboratory exploitation, their bodies became sites for the expression of fear of the uncomfortable entanglement of bodies across human/animal and first/third-world boundaries in the emerging pharmaceutical imperialism of mid-century. If scientific medicine relied on vivisected animals for the production of pharmaceuticals, then public health officials could turn away from militarized forms of spatial eradication to focus on high-tech medicines, adapting the body's internal systems rather than its environment. Although this development was promising because it decreased the need for quarantine and other invasive or carceral territorial approaches to disease eradication, public fears proliferated concerning vivisection, the use of human and nonhuman bodies for invasive, painful, and potentially deadly laboratory or clinical experimentation.

The resulting linkage of fear and hope in rapid medical change from World War II through the early Cold War era connected public representations of rhesus, chimpanzee, and gorilla bodies to anxieties over the mobility and adaptability of the national body. This related to concerns about whether US empire could contain the twin geopolitical forces of communism and decolonization. I trace these links through both scientists' accounts of the development of national primate research institutions during the polio scare and films and other images depicting mad scientists, medical experimentation, spinal taps, and inoculation in polio research. In the process, I argue in this chapter that polio virus gave particular form to national anxieties as new pharmaceuticals, the changing geopolitical role of the United States, and relations between humans, monkeys, and microbes were shifting the biopolitics of empire. Even as more animals were commoditized for medical research and production, these intersections paradoxically helped make imaginable a space of interspecies feeling between humans and nonhuman primates. Fear of polio was channeled into a vision of interspecies kinship, one betrayed by (yet entangled with) the cold fact of mass exploitation of human populations and nonhuman primates in scientific medicine. The entrenchment of primate vivisection in the imperial architectures of disease control produced new configurations of the government of species that are crucial to understanding the so-called antibiotic revolution of midcentury.

Monkey Grinders: Epidemic Polio, Animal Modeling, and the Paralytic Spine

Medical historians often note the optimism of the turn to scientific medicine and the development of wonder drugs in the United States in the middle of the twentieth century, a time when the technological developments of World War II and the early Cold War era generated both an exceptionalist discourse of American technological supremacy and fears of rapid technological change. The 1940s and 1950s saw the wide dissemination of penicillin (an antibiotic drug that worked broadly against bacterial infections), polio vaccine (the first effective vaccine to be developed since the smallpox vaccine in the 1790s), and gamma globulin (a blood compound that effectively stimulated immune response). Each of these pharmaceutical products contributed to changes in the forms of intervention used to combat disease.[6] In the so-called antibiotic revolution of midcentury, biomedical consumers could increasingly control the risks of the outside world by consuming targeted or generic pharmaceutical commodities. This was a preventative or prophylactic regime of disease control that encouraged pharmaceutical consumers to manage viral and bacterial risks by optimizing the time of internal immune response rather than the space of environmental contact with infectious agents. By exposing bodies to dead or contained antigens in vaccines, the immune system could be primed for infection in advance. In the event that infection occurred, the immune response could itself be amplified or supplemented through other drugs. This shift generated new forms of optimism among patients, practitioners, and medical researchers that medical science would one day win total defeat over microbial vectors. Doctors began to widely prescribe penicillin, and researchers expressed faith in "the ability of the pharmaceutical industry to supply an increasing range of targeted and effective remedies."[7] As David Serlin has written, the emergence of a medical consumer culture in the 1950s helped make US publics understand drugs and other medical technologies as essential components of "a prosperous economy and a modern social self."[8]

At this time, public health discourses began to shift away from an emphasis on spatially defending the body and toward regulating the internal immune response. Emily Martin gives the example of a 1955 image from *Life* magazine depicting the "lilliputian hordes" of bacteria and viruses entering a prone white body. By the 1940s, mainstream reporting on diseases began to stress the importance of proper internal maintenance of the body and the cohesion of the skin's physical barrier in order to prevent the transmission of disease. The idea that the generic antibody would defend the human organism internally became a dominant component of popular conceptions of health.[9] Disease control would be as much about the body's reproductive process for engineering immunity as it would be about maintaining sanitized spaces.

Polio vaccine played an important role in these shifts. After decades of research, the production of effective vaccines by Jonas Salk in 1955 and Albert Sabin in 1961 became cause for expecting that biomedical research could one day cure the entire gamut of infectious diseases. "The excitement, rejoicing, and confusion about the Salk polio vaccine," according to an article in *Life* magazine, had overshadowed the coming "larger victory" against "man's last intractable enemies in the world of germs, the viruses."[10] Although this optimism generated by polio vaccine is often emphasized in histories of midcentury transformations in health and medicine, it is important to remember that recurrent polio epidemics in the United States during the first half of the twentieth century made paralytic polio impacting the central nervous system (then called infantile paralysis) a dreaded disease, with epidemics intensifying during the polio scare of 1948–55. Despite the fact that polio was less pervasive than scarlet fever or tuberculosis, the crippling nature of the disease, its devastating effects on children, and the relative lack of understanding of its modes of transmission and progression provoked public anxiety. Attempts to control polio targeted the working classes and immigrant groups despite the fact that the disease often impacted the young, wealthy, and white and those living in conditions understood to be sanitary. Furthermore, wide disparities of access to health care and rehabilitation also worked to sow fears of the disease, as the Depression and a lack of institutions in many parts of the country made ongoing care impossible for large portions of the population.[11] There was thus intense public concern over the disease from the Depression through World War II, a period before private philanthropies subsidized individual therapy and during which the experience of treatment appeared highly experimental, as "most polio patients were treated locally by whatever orthodox or unorthodox means available."[12]

In this context, there was a particular site on the body that animal vivisection made central to public visions of changing forms of disease intervention: the spinal column. If Mieth's photographic gaze relied on the primate nervous system's deep subjective connection of emotion and motion, polio researchers in the early twentieth century attempted to engage in the battle for immunity by entering human and monkey nervous systems more directly through the needle and the scalpel. Polio research intervened directly and painfully in the body through the diagnostic and experimental procedure of the spinal tap, also called lumbar puncture. The spinal tap involves inserting a long needle between lumbar vertebrae in the middle of the back and pressing past the dura mater, where the needle can extract cerebrospinal fluid for diagnostic examination or, during the era of polio experimentation, for the development of vaccine serum. This pathway into regulating the immune response of fluids that make immediate contact with the nerves was a crucial site of medical research given the early twentieth-century hypothesis that virus accumulated in the spinal column might be related to polio's most feared pathological transition: paralysis.

Key European and US American polio researchers including Simon Flexner, Karl Landsteiner, and Constantin Levaditi adopted rhesus macaques as an experimental model. Because they were unsure of the pathway and progression of the disease, researchers often focused on its most dangerous effects located in the nervous system. Having developed effective tests for polio using lumbar punctures, researchers attempted to demonstrate that the poliovirus present in the cerebrospinal fluid caused the disease. In 1908, Landsteiner and Edwin Popper demonstrated the cause of paralysis through spinal infection with the poliovirus. After failing to transfer the disease to mice, guinea pigs, and rabbits, they injected the ground spinal cord matter of an Austrian child who had died of polio into the body cavities of one baboon and one rhesus monkey.[13] Although this experiment produced paralysis in both subjects and generalized symptoms of polio in the rhesus, this was an artificial adaptation of the virus that did not accurately model the disease pathway for humans.[14] The most severe forms of polio impacting the human nervous system emerged from an oral pathway. Ingested polio first colonizes the gastrointestinal system, where immune response normally contains the virus, producing no symptoms. In cases where the virus spreads to the bloodstream, influenza-like symptoms emerge across the body. In a small percentage of cases, virus penetrates the spinal cord and/or the brain stem, impairing motor function and leading to the potentially severe debilities of paralytic polio of the central nervous system that were glimpsed by the researchers' discovery of virus in the cerebrospinal fluid. But the rhesus model of the disease in the laboratory bypassed these steps of transmission, in effect inventing a new, direct pathway for paralytic polio in monkeys.

In 1910 and 1916, Simon Flexner, Paul A. Lewis, and Harold L. Amoss published a series of medical articles focusing on the presence of poliovirus in the lymph and cerebrospinal fluid and testing serum therapy through inoculation of spinal material. Writing for the Rockefeller Institute during the New York polio epidemic of 1916, Flexner and Amoss warn of the widespread prevalence of asymptomatic carriers of poliovirus and suggest that minute variability in the permeability of lymph and cerebrospinal fluid vessels allows the virus to migrate from the respiratory system directly into the nervous system. Flexner and Amoss developed a gruesome experimental method drawing on Landsteiner and Popper's original use of ground infected spinal cord. They experimented with the transplantation of active virus suspended in rhesus cerebrospinal fluid and centrifuged spinal cord and medulla matter, demonstrating the potential to transfer the infection from ill to healthy monkeys. Then they moved on to inject the cerebrospinal fluid and centrifuged spinal matter of immune monkeys. Reporting no signs of infection following such inoculation, they report, "This experiment is conclusive in that it indicates that an immune serum alone . . . prevents the infection. The experiment has also a bearing on the serum therapy of human poliomyelitis and upon the question of the employment for intraspinal injection of normal sera and other fluids."[15]

This statement of hope for a cure derived from the rhesus body made demand for these animals high. Rhesus bodies were consolidated as the experimental model and their spines turned into the prized and costly source for experimental cures for paralysis. The literal harvesting of the rhesus nervous system appeared as the best scientific route to defending the human nervous system and its capacity for bodily mobility from the scourge of polio. Even as rhesus became key resources for this scientific approach to polio, children and prisoners were also made into experimental subjects on the other end of the process of drug development: clinical trials of vaccine sera. These trials in the 1930s and the 1950s famously included the use of prisoners as well as both institutionalized and healthy children (1.8 million US children were included in the trial for the Salk vaccine). The experimental vaccines produced for these trials came from culturing virus using bodily fluids and tissues from monkeys, and occasionally from mice, horses, and other mammalian species. Polio research at midcentury involved the engineering of national immunity by dispersing death across primate species and dispersing pharmaceutical risk across precarious test populations who were incarcerated, infected, or at risk of infection. Human and animal experimentation thus involved varied forms of state force against bodies, dispersed through complex geographies that transformed animals into capital and human subjects into experimental medical citizens whose compromised consent could be drawn into research. In such ways, bodies drawn into varied forms of laboratory and clinical knowledge production were biopolitically segmented into increasingly complex configurations of social and biological precarity.

Polio and the Cold War Primate Trade

Although histories of the polio scare and the invention of polio vaccines are often presented in a national framework detailing US medical, state, and philanthropic efforts to conquer an intractable contagion, the dependence of polio research on the trade in Asian-origin primates deeply entangles the story of polio with the histories of decolonization and the Cold War. Rhesus macaques are Old World monkeys indigenous to northern India (where they are often considered pests). Like most large primates, they are considered anatomically and genetically similar to humans, a presumed benefit for vaccine experimentation. In the course of US and European polio research and vaccine production, hundreds of thousands of rhesus macaques were captured and exported by Indian trapping operations. Yet the importation of these experimental animals became increasingly difficult as demand grew to unprecedented levels. In 1936, William Park and Maurice Brodie carried out vaccine trials on 11,000 human subjects (mostly children) using serum from ground rhesus spinal matter. Soon after, John Kolmer used live virus to test vaccine on 10,000 children. After several of Kolmer's subjects contracted polio and allegations circulated that the same may have happened in the poorly documented Park-Brodie trials, public outcry stalled experimental testing of vaccine until the 1950s.[16] Later, during the Salk vaccine trials, the production source was shifted to rhesus kidney cells and, eventually, to cell culture methods that significantly reduced demand for rhesus. Yet through the height of the polio scare and of mass vaccine production in the 1950s, rhesus remained key models for polio research and sources of serum production. Rhesus were thus both raw materials and laborers for scientific knowledge production. Rhesus bodies could be literally harvested as vaccine medium, even as their reproductive processes were engineered as models to demonstrate the transition states of disease and the efficacy of vaccine therapies.

Because of the early successes in polio research, rhesus monkeys soon became prized research subjects for other medical specialties. After the end of the war, the justification for expanding importation and breeding of rhesus was based on expected demand in both polio vaccine production and in fields including aging, radiation, psychology, heart disease, and blindness. Yet even after the scare over contamination of the Salk vaccine with simian virus led to a shift away from the use of rhesus macaques for vaccine in the early 1960s, US health officials saw rhesus as central to the future of scientific medicine. Thus from 1939 to 1961, US researchers and public health officials designed a national system for the importation and breeding of nonhuman primates, which were for the first time designated as vital national "research resources."[17]

It was Columbia University comparative anatomist Clarence Ray Carpenter who first attempted mass primate importation schemes. In 1939, Carpenter transported 409 rhesus macaques from Calcutta to Cayo Santiago, the islet off the southeast coast of Puerto Rico where Mieth took her famous photograph for *Life* magazine (figure 3.2). Carpenter convinced Columbia to purchase Cayo Santiago as part of its larger plans for a tropical medicine institute. Even though Carpenter himself was not a lab researcher — he did noninvasive behavioral research on large troops of free-ranging primates — he justified his scheme to import hundreds of Indian-origin monkeys to US-occupied Puerto

Figure 3.2 Aerial photograph of Cayo Santiago, 1938. Penn State University Archives, Pennsylvania State University Libraries.

Rico with reference to infectious disease research, apparently alarming residents of nearby Punta Santiago about the potential for zoonotic outbreaks.[18] Despite this attempt to align the monkey scheme with the modern laboratory and the national defense, narratives of Carpenter's importation scheme also stress the monkey's backward native social and ecological settings in India. *Life* magazine announced the 1939 launch of Cayo Santiago by invoking familiar orientalist tropes concerning the monkeys' Indian origins: "Because he is considered sacred in India," claimed the unsigned article, "the rhesus is domineering, undisciplined and bad tempered."[19] Nonetheless, importation is necessary "because Mahatma Gandhi is preaching against the exportation of the sacred rhesus monkey."[20] In a 1940 article in the journal *Science*, Carpenter denounces the exuberance with which the Society for the Prevention of Cruelty to Animals, Hindus, Buddhists, conservationists, and even the British colonial state attempted to limit the primate trade. Monkeys were a "necessary import," particularly for infectious disease research. Carpenter implies that corrupt nationalists manipulated Indians to support animal protection: "These peoples . . . are told that monkeys are used for the 'rejuvenation of decadent Westerners.'"[21] In response, Carpenter suggests that strong US diplomacy with the British Raj would secure scientists a steady supply of Indian monkeys.

Carpenter's first attempt at nationalizing a self-renewing population of rhesus reflects a colonial logic, assuming that as rhesus were imported into the United States, they needed to be maintained in an island habitat replicating an idealized view of tropical nature.[22] Perhaps influenced by media visions of US-occupied Caribbean and Pacific islands filled with palms and coconuts, Columbia officials tasked with landscaping Cayo Santiago transformed its pastureland into an idyllic tropical landscape. The New Deal Civilian Conservation Corps was enlisted to forest the island with mahogany, coconut palms, fruit trees, and root vegetables. In advance of the release of the monkeys, the *Illustrated London News* overlooked the islet's complex history involving the sugar trade and its use as sheep pasture to present it as *terra nullius* augmented slightly to feed its primate settlers: "It is . . . rocky land covered with shade-trees and thickets; some of it sandy depression, fringed with coconut-palms, and some flat, earthy expanses in which . . . root crops have been planted for the future islanders. . . . It is, in effect, a miniature Pitcairn Island dropped into a shallow sea of parrot-wing blue, jade green, or wrinkled copper, according to the caprice of the sun and wind."[23] Yet save for a few coconut palms, most of the transplanted organisms were killed off by climate or unruly monkeys.[24] Presuming that tropical fruit and root vegetables would indefinitely nourish animals and that the sea would enclose them, caretakers were surprised when the monkeys destroyed the expensive fruit trees and, in small numbers, swam half a mile to the Puerto Rican main (figure 3.3). Tearing down the imperialist fantasy of tropical nature, the animals forced scientists to establish feeding stations. Cayo Santiago ultimately had to be maintained as a "semi-free ranging" site, with feeding stations featuring Purina monkey food and cages for intermittent examinations becoming standard parts of life on the island.

Despite the effort put into establishing the field station, persistent tuberculosis infections among the rhesus, institutional disputes, and funding difficulties plagued Cayo Santiago in the decade following the war. These pressures became especially acute as the University of Puerto Rico sought increased control of the Columbia School of Tropical Medicine in order to promote a broad public health agenda instead of the mercurial research agendas of Carpenter and others from the US mainland. This assertion of Puerto Rican control of the institution, occurring within a broader rise of nationalist politics in Puerto

Figure 3.3 View of Cayo Santiago from Puerto Rican mainland, from *Illustrated London News*, August 13, 1938. Courtesy Special Collections, Conrado F. Asenjo Library, Medical Sciences Campus, University of Puerto Rico.

Rico, coincided with Carpenter's abandonment of Cayo Santiago after completing his two years of study there. In the decade that followed, there was little interest in pursuing similar rhesus schemes.

Yet in 1947, just before *Time* magazine would advertise the onset of the polio scare with tens of thousands of US infections annually, India finally attained national independence from Britain. As pharmaceutical companies, hospitals, and labs sought to import rhesus directly from India, they encountered new demands from the postindependence state concerning the treatment of animals and the potential conflicts between security research and India's position of nonalignment with respect to the Cold War battle between the United States and Soviets. In 1955, noting the spike in rhesus demand with the Salk vaccine, the Indian government placed a brief moratorium on rhesus exports to the United States. Following publicity of the US Department of Defense's wartime nuclear weapons research on chimpanzees and other primates, India negotiated an agreement that would resume exports on the condition that Indian primates would not be used in radiological experiments. By the 1970s, several African and Asian countries followed this precedent and imposed primate export bans in line with Prime Minister Jawaharlal Nehru's lead in the non-aligned movement's push for nuclear disarmament.[25] Even though the United States was able to reestablish access with new assurances that Indian animals would not be used in radiological testing, public health officials emphasized the decolonial threat to research: "The problem of supplying monkeys is now acute and there is no assurance whatever that it will improve. The immediate reason is the necessity for poliomyelitis vaccine, but in the recent past an embargo on Indian monkeys was nearly catastrophic."[26]

Noting the contrast between heavy Soviet investment in primate models for heart research and the underfunded colony at Cayo Santiago, several officials at the National Institutes of Health (NIH) as well as in the Department of Defense and the State Department began discussions about expanding Carpenter's original primate colony idea to a multisited primate research network.[27] Tulane University cardiologist George A. Burch convened the Committee for the Establishment of a National Cardiovascular Primate Colony at NIH. From 1956 to 1963, this committee would bring about the development of the national system for biomedical primate importation and breeding known as the Regional Primate Research Centers.[28] These institutions were established as domestic centers only following a lengthy process in which species selection, locations, architectures of confinement, decolonial political entanglements, and public marketing of the institutions were discussed at length. Balancing the emphasis on Indian rhesus imports with the potential for other species from different geographic locations to be incorporated into research, the committee settled on a domestic, largely indoor operation only after lengthy discussion of different options. There was initially no consensus on how best to supply and house researchers with primates. Should scientists rely primarily on importation of foreign animals, where breeding occurs on the free range, but where control over the habitat and stocks was in the hands of other countries with other interests? Should the state pursue domestic breeding colonies in free-ranging settings that simulate primates' natural environments, as at Cayo Santiago? Or would it be workable to undertake monitored breeding indoors? With disease and genetic variations proving to be the main obstacles to the effective use of rhesus monkeys in research, what institutional organization could provide for the health and well-being of increasingly standardized populations of rhesus or other appropriate species? Would the state be able to defend monkeys not just from disease, but also from the recurring threats of war, underfunding, and political critiques of research? Over seven years, the committee's operations moved from Carpenter's visions of outdoor tropical primate colonies to indoor captive institutions that could domesticate the rhesus macaque in three senses: rhesus macaques were literally imported into the continental boundaries of the nation; they were broken of their cultural association with a wild and untamed animality; and they were increasingly held captive in indoor labs that attempted to establish their suitability to model human bodies and minds.

Rhesus imports during the mid-1950s were the highest ever, numbering around 150,000 annually, with 120,000 killed for polio vaccine production.[29] At this moment, research officials noted a crisis of rhesus resources caused by both high demand and problems of Indian supply. Paul Weiss, a National Research Council official, claimed, "Conservation . . . is not really the only answer, but what we have to have is a balance sheet between production and consumption" of experimental research primates.[30] With nearly 90 percent of imported monkeys used for polio vaccine production, researchers noted the increasing difficulty in securing imports for small research projects and for specified types of animals.[31] Leon Schmidt, a researcher at a Cincinnati hospital, claimed, "We have passed the point of availability of the larger animals and we're dipping into the young now who would be under normal circumstance the creators of the next generation. . . . We are in a very, very precarious situation with respect to procurement of rhesus monkeys from the particular areas of India where they have been found in former years."[32] Solving this problem was viewed as a national priority on par with provision of other key raw materials of the security state: "The rhesus monkey is almost as strategic material as tungsten and tin and natural rubber. . . . We may do to the rhesus what we've almost done to the buffalo here in the United States. It would seem therefore, that any device which would bring together people who could make maximum national use of any of the animals would be a very worthwhile procedure."[33]

Given the long colonial association of primate evolution with Africa, it is perhaps no surprise that the committee explored establishing US-sponsored free-ranging colonies there. Two research sites were considered. The first, approved by the outgoing Belgian Congo government in 1959, would use US funds to expand an existing Belgian-run research institution with a new field station for capture and breeding of chimpanzees on a river island in the vicinity of Stanleyville (present-day Kisangani). Envisioning a free-ranging site that would allow biomedical harvesting of chimpanzees, US and Belgian officials entered into detailed plans for expanding the Institution for Scientific Research in Central Africa. Yet in 1960 and 1961, the dramatic events of Congo independence thwarted Belgian researchers' attempts to maintain their colonial institute, leading eventually to United Nations assistance in airlifting chimpanzees away. The second proposed venture, the Darjani Primate Center near Kibwezi, Kenya, would house a variety of primate species and lure US researchers to visit through the promise of safari adventures. Nonetheless, the station never housed more than a handful of primate subjects, and after Kenyan independence, most researchers relocated to the capital.

The demise of the African institutions at the moment of Kenyan and Congolese independence in the early 1960s meant that the NIH committee's focus returned to the framework of Indian importation, to be supplemented by domestic breeding. Nationalist politics in Puerto Rico, decolonization in India and central Africa, and the rise of a third world nonaligned movement all worked to constrain the unfettered neocolonial expropriation of primate subjects by the US security state, encouraging scientists to domesticate primates and to take on the project of managing their reproduction and standardization.

These geopolitical developments led scientists, in turn, to think about the psychological and biological needs of animals whose populations were now to be maintained domestically. If the directly imported rhesus was treated mainly as an exploitable commodity for biological production, the domestically bred rhesus needed to be maintained in a form of captivity that allowed for smooth reproduction and exploitation. This led to discussions of the proper treatment of monkeys that exceeded simple logics of science or profit; researchers began to think about animals as intentional beings. In committee meetings, a major controversy centered on whether keeping monkeys in captivity went against proper animal care. Harry Harlow and Leon Schmidt argued in favor of indoor facilities, speaking of Harlow's success maintaining rhesus in "very small quarters." By contrast, Irving Wright, a Cornell cardiologist, said, "[I am] not quite sure that locking monkeys up in cages is really giving them a natural environment, if you are going to study their long-term and generation after generation effect on them. . . . I wouldn't expect it to be true of humans; and I don't see why it would be true of monkeys." George Burch, signaling the potential effect of confinement on the body, added that "it might change their personalities," with Wright implying further that captivity might "affect their outlook on life." Noting the benefits of representing both approaches within the system, James Watt, former researcher at Cayo Santiago and director of the National Heart Institute, argued for including different living arrangements within the system. When "talking about a colony," Watt said, researchers are "shooting for . . . a diverse group of workers."[34]

The NIH program ultimately established a mix of indoor and outdoor facilities, with large numbers of rhesus eventually bred and housed in highly controlled indoor settings. This arrangement was largely indebted to University of Wisconsin comparative psychologist Harlow, who had recently demonstrated the efficacy of indoor captive breeding and study. Harlow's work— before and after the development of the primate centers—helped break the rhesus of its association with tropical nature, making the macaque a domesticated figure of national progress.[35] Depriving rhesus of social interaction and kin affection, Harlow attempted to prove the evolutionary significance of mother love by brutally withholding it from juvenile rhesus.[36] His labs were also successful breeding facilities where he developed isolation techniques for ensuring disease-free (specifically tuberculosis-free) monkeys that he eventually sold to other researchers.[37] In bringing the rhesus indoors, establishing an ideologically loaded relation of the macaque to the mother, emphasizing monkeys' affective life and complexity, and publicizing universal psychological characteristics shared with humans, Harlow worked to establish a new vision of the rhesus divorced from the tropical habitat, the mad scientist, and colonial expeditions. Making the laboratory into a home, the rhesus was rendered domestic—cut off from the transnational circuits of importation, placed under the familial care of a mother, and removed from the old associations of mimicry and sideshow performance. Even as polio spurred the mass vivisection of rhesus macaques and other species of nonhuman primates for infectious disease research, decolonization required that this project not only develop as a necropolitical exploitation of vivisected and commoditized animal capital, but also a project in reproduction and affection, a lively attempt to attach rhesus development and socialization to the spatial logics of captivity. Debates regarding the proper domestic environments of the experimental macaque include moments in which rhesus are viewed not simply as resources, but also as intentional beings with complex minds and needs. Here we have the transcolonial roots of medical science's ambivalent production of the nonhuman primate as almost, but not quite, human.

"Life in That Tiny Bottle!": Inoculation, Kinesis, and *The Ape*

The work of primate researchers to domesticate the Indian rhesus macaque was intimately related to shifts in how rhesus macaques and medical researchers were portrayed in popular visual media. Mieth's photograph is an early example of the potential that medical science established for such a shift; even though the monkey retained a trace of colonial sensationalism, linked to a caricature of Hindu patriarchy, it could nonetheless be integrated into visions of national domesticity and the shared capacities of primate nervous systems for movement and feeling. Yet Mieth's image was published at a time when there remained deep public skepticism of the aggressive vivisection of the research scientist, particularly after the failure of the 1936 polio vaccine trials that both utilized ground rhesus spinal matter and endangered the white child. As such, even as rhesus were undergoing multiple processes of domestication, they were also deeply associated with public anxieties concerning medical intervention into the national body.

Hollywood horror films during World War II publicized anxieties over changes in US Americans' relationships to technology, medicine, and politics, mobilizing images of spinal taps, mysterious injections, xenotransplantation, and paralysis and exploring the potentials for injury and death in laboratory and clinical research. In these films, apes and monkeys were not simply raw materials serving the interests of science. They were also vivisected bodies, victims, tricksters, violent beasts, and vampires, threatening delicate social norms that seemed at risk given changes wrought by modern war, transit, and technology. The experimental primate—imported from the colonial jungle in order to burnish national immunities—appeared as both a potential savior and a potentially uncontainable risk of an era of global warfare. The simultaneous linkage of the research primate to technological optimism and the corruption of nature was dramatized within the romanticized domestic spaces of the small town and the home. When horror films directly represented polio, whiteness took on heightened significance, as viral contagion and the new technocracy of biomedical research erupted into the fabric of segregated white communities. These communities faced both the disease's threat to the species and biomedical technology as the nation's emergent yet unfamiliar form of security. In response to a regime of biopower in which the settler population was disciplined into viewing bodies as optimizable through experimental therapies, anxieties over biomedical progress generated monstrous spectacles of interspecies intimacy and techno-vampirism.

In this section, I explore depictions of the experimental laboratory primate, the mad scientist, and the polio patient in the 1940 film *The Ape*. In many ways, this film, directed by William Nigh and starring prolific B-movie actor Boris Karloff of *Frankenstein* (1931), is unremarkable—it is one of a number of low-budget wartime horror films following formulaic plotlines suggesting the corruption of modern life through disease, criminality, and technological change. Yet *The Ape* is remarkable for its attention to polio and vivisection in the spheres of the home and the small town, even as it suggests that these conventionally whitened spaces were constituted within a broader world of disease, debility, and death against which the life of the community had to be defended. In the film, Karloff's character, Dr. Bernard Adrian, is the only doctor in Red Creek, the rural town depicted in the film. Having been expelled from a philanthropic research institution for unethical experiments, Adrian moved to the town during a prior epidemic of paralysis. Feared by the townspeople because of rumors of his experimentation on humans and dogs, Dr. Adrian has one regular patient, his single defender in public: Frances (played by Maris Wrixon), a twenty-something, wheelchair-bound paraplegic polio patient. When an escaped circus gorilla strangles its trainer, crushing the man's upper vertebrae, Dr. Adrian harvests the man's spinal fluid to provide Frances with an experimental cure, attempting to restore motor control of her paralyzed legs. In the pre-Salk-vaccine world of the film, the serum from the corpse is imagined as successful in actually restoring movement rather than simply preventing the spread of infection. The serum works to restore the movement of Frances's legs, but she requires more healthy spinal fluid to walk. Taking advantage of the town paranoia over the missing gorilla, the desperate doctor dons a costume to masquerade as the terrifying ape, setting out to kill locals in search of more spinal fluid.

Dr. Adrian emerges as the conflicted character at the heart of the intervention of modern biomedicine into life itself. The portrayal of Dr. Adrian as mad scientist unfolds by attributing to him excessive empathy and ambition through his relationship to the patient, Frances. The campaigns of the March of Dimes had already deeply sentimentalized the wheelchair-bound polio patient, the poster child whose tragic disability required the vigilant participation of the public in raising money for the cure.[38] Frances is the sentimental figure driving the plot; her wheelchair-bound immobility mobilizes the desires of both Dr. Adrian and Frances's love interest, Danny. Frances's paralyzed legs remain out of sight and under a blanket, hiding her impairment. Frances embodies the threat of polio to white femininity, as she reminds the doctor of his own daughter who succumbed to polio during the earlier epidemic. The substitutability of these bodies and Dr. Adrian's melancholic attachment to Frances emerge through the doctor's presentation of a gift, the jewelry of his deceased daughter. Drawing Frances into the sphere of kin, the doctor's gift figures her body as a site for restoring the nuclear family endangered by polio. Yet the doctor's quick turn to military metaphors against paralysis — the talk of medical weapons deemed "frightening" by Frances—unveils the empathetic doctor's will to power. The film treads a fine line here: even as Dr. Adrian is the subject of excessive affect produced by loss, the film also rejects the townspeople's ostracization of him as rooted in a self-destructive failure to embrace the potential progress of modern science. Yet Adrian's excessive ambition and appropriation of nature's power over life also reveals a disturbing violence behind medical technology, which propels the suspense of the narrative as Adrian desperately seeks out a way to reproduce the cure.

The horror of the medical researcher's will to power inheres in the imagery of the spinal tap, the breaching of the skin of the body to extract life itself and, in the process, to kill. A circus trainer injured by the escaped gorilla becomes the site of the doctor's appropriation of the power to give and take life. Brought to the doctor for treatment after the attack, the dying man becomes Adrian's source for an experimental polio treatment he hopes to harvest through a spinal tap. As the man pleads from the exam table, "Don't let me die," Adrian muses over his serendipitous opportunity to obtain human serum: "Man—the highest animal." The last words he says to the man before death make clear the doctor's ambition, couched in the languages of social progress and personal redemption: "I'm going to make you medical history. I'm going to keep a promise," the cure promised to Frances and presumably Adrian's daughter as well. Displaying the vial of cerebrospinal fluid following the trainer's death, Adrian exclaims, "I've succeeded! It's there—life. Life in that tiny bottle!" (figure 3.4). The medical narrative here is not expressed in the language of modern immunology; it is rather invested in the sacred relation between life and the body. There is no specific agent within the body whose power is harnessed by Adrian to fight a specific disease; rather, the doctor describes his extraction as life itself.

The terror of invasive procedures such as spinal taps and hormone extractions became a common trope of the perils of biomedical research in the 1940s. These procedures not only breached the border of the skin but also suggested a disturbing interpenetration of bodies that called into question both the independence of the liberal-humanist subject and the purities of race and species. This is perhaps best exemplified by the vampirism of spinal taps in 1943 film *The Ape Man*, in which Bela Lugosi of *Dracula* (1931) plays a scientist whose attempt at ape-to-human spinal xenotransplantation goes awry in his home laboratory. He is transformed into a fang-bearing human-ape hybrid, whose excessive animal affect is expressed through criminal acts. As he claims to feel himself "filling with the animal instinct," he desperately sets out to

Figure 3.4 Dr. Bernard Adrian (played by Boris Karloff) holds "life in that tiny bottle!" in *The Ape* (1940), directed by William Nigh.

commit several murders to harvest human spinal fluid in a last-ditch attempt to rehumanize himself. In *The Ape Man*, the mad scientist unites the affective power of two monsters: the vampire, an index of technocratic power, and the ape, an index of the primitive violence underlying and enabling social change.

In the eyes of *The Ape*'s Dr. Adrian, the polio patient Frances is a figure for the broader promise of scientific medicine. When Danny protests against the lack of information on the experimental treatment as well as the pain involved in restoring feeling and movement to Frances's legs, Dr. Adrian begins to tell Danny that "through her" the treatment will "rid the whole world—" but Danny interrupts. In a tense moment, Danny protests to Adrian: "I love her just the way she is, Doc! I can take care of her—I want to take care of her! I don't want no experimenting on her." Danny is "not in love with the world" but "in love with her." Frances exists on the threshold of accepting the technological promise of engineered immunity that threatens to distance her from the moral community, to compromise her positions as white/human/woman within the Jim Crow settler society. The resolution to this tension lies in the distinction between the two ethics of care offered by the doctor and

Danny. The boyfriend envisions a long-term caregiver relationship—one of dependency of the female polio patient on the companionate couple form and the man as economic provider. In contrast, the doctor offers a liberal promise of future independence—the restoration of the embodied capacity of the human, and thus the nation through her labor as patient. This requires a reciprocal relationship—there is not just the doctor fighting against the disease, there is also his stoic conscript, the patient who must subject herself to experimental forms of immune optimization. Throughout the film, the patient is subjected to tough love. After her initial moment of fright at Dr. Adrian's pronouncement of a cure, Frances is presented as a stoic patient, confined to her wheelchair but always well dressed, clean, and, to the extent possible, erect. She labors to embody the cure in contrast to the tragic figure of the polio poster child.

In his pursuit of the hypermodern cure of pharmaceutical empire, Dr. Adrian becomes the mad scientist, himself "going ape." The escaped circus gorilla named Abu eventually attempts to break into Adrian's laboratory. The moment of the ape's violent entrance into the lab—ostensibly pursuing the cruel trainer from whom Adrian extracted his polio serum—brings the touch of unruly nature, the primitive, the jungle, which threatens the promise of a disciplined medical modernity (figure 3.5). In the heat of the ape's attack, Adrian stabs the animal, then quickly realizes that the town's fear of the escaped animal can work to his advantage. In order to obtain more serum necessary for Frances's cure, Dr. Adrian does something remarkable: he skins the gorilla and wears the remains, successfully passing as a gorilla. He then sets out to attack a local policeman, hoping to harvest more spinal serum. The policeman, however, stabs gorilla-Adrian and summons other officers who eventually shoot the costumed doctor on the front steps of his home clinic and laboratory.

Although Dr. Adrian expresses deep concern for Frances throughout the film—always bordering on the erotic and bringing him into conflict with Danny—his relation to Frances is complicated at the moment he dons the ape suit. Michael Lundblad reminds us that apes came to spectacularly represent the myth of the black male rapist in early twentieth-century US literature, while Megan Glick asserts that the gorilla's apparent blackness was contrasted with the whiteness attributed to chimpanzees, seen as the highest of the nonhuman primates.[39] The respective association of Asianness, blackness, and whiteness with the rhesus, gorilla, and chimpanzee signals a racial triangulation through which the government of species mediated between the poten-

Figure 3.5 Circus gorilla Abu breaks into Dr. Adrian's lab in *The Ape* (1940), directed by William Nigh.

tials of containment and contagion.[40] Thus the film's substitution of the Indian rhesus from the polio lab with the African gorilla of the jungle and the sideshow may reflect a precise racial schema underlying the film's fear of vivisection rather than an unscientific conflation of ape and monkey. The doctor's injection of his "life" serum into Frances makes the racialized association of medicine with sexual penetration explicit, as does the repeated use of the language of the lynch mob by the townspeople who complain about the doctor. If the patient can become an experimental subject, then the embrace of medical technology risks threatening the heteronormative structures of domesticity through the interpenetration of bodies, races, species, and genders.

Despite this, the climax of the film conjoins fear of medical intervention to hope for a cure. Once the police shoot gorilla-Adrian as he attempts to return home, a sobbing Frances—in her first moment of self-propelled motion in the film—wheels herself up to the front steps to see the dying doctor. After the police remove the gorilla suit, Frances approaches the doctor, displaying her own motion as the fruit of his labor. She dramatically stands, displaying her newfound ability to, however hesitantly, take control of motion (figure 3.6). Visually staging the now towering Frances on the side of the town, opposed in the shot to the fallen Dr. Adrian and the gorilla skin, this scene returns

Figure 3.6 Polio patient Frances (played by Maris Wrixon) rises from her wheelchair in *The Ape* (1940), directed by William Nigh.

Frances to the community that has spurned the doctor in the moment it finally recognizes medical science's powers of transformation. As the film reverses the positions of able-bodied and disabled, kinesis and paralysis, Dr. Adrian and the ape are joined as sacrificial figures, destroyed by the violent promise of modern medicine. Frances's reintegration into the social sphere of the town is dependent on the violence of the laboratory as well as the sacrifices of the trainer, the animal, and the doctor. The immense technological promise of pharmaceutical empire is catalyzed in the literal incorporation of sacrificed lives into the cure.

Much of the complexity of Nigh's *The Ape* lies in its sensational motif of sacrifice, in which the horror of medicine resides in the newfound vital transference and substitutability that can cross between human bodies as well as the mythic borders separating species. This is portrayed visually through the practices of spinal tap and inoculation and in the transformation of Frances from a figure of paralysis to one of kinesis. This is a transition that, following Muybridge's use of animals to turn the still photograph into the moving picture, is constitutive of film as a visual medium. In addition to depicting animal motion, paralytic polio and other debilities were central to producing the visual wonder of a lengthy list of early films.[41] Whereas Frances earlier had mused from her wheelchair at the movements of circus acrobats who displayed an ideal "muscular grace and coordination," the film wraps up with an abrupt happy ending in which Frances assures Danny that she will never again be confined to a wheelchair. Overcoming debility, then, is premised on the expansion of certain bodies rendered through the nonconsensual appropriation of others (including the working-class trainer and the experimental primate). This biopolitical segmentation can be visualized but cast aside with the possible restoration of able-bodied white femininity to the heterosexual couple and, thus, to the everyday norms of small-town life.

Nigh's film expresses an awareness of unequal access to biomedicine's engineered immunity. At least three human men die in the film, two of them Adrian's patients, and one of whom is a working-class man whose life goes unprotected by the town. Thus the film visualizes trade-offs in which medicine is mobilized for certain privileged objects of security over others. These logics of ontological ordering are evident in the organization of social space in the film. Like the doctor—himself a liminal figure in the community—the ape, the trainer, and the circus that brings them to Red Creek appear on the outskirts of town, brought in to interrupt its normal functioning by the train system that signals the perils and promise that come from increased integration into national and international circuits of transit and exchange. As such, both the ape and the polio patient become figural representations of the differential mobilities of military and economic globalization during World War II. This is apparent in the ending of the film, which enacts the predictable convention of narrative prosthesis described earlier in chapter 1. Mitchell and Snyder describe narrative prosthesis as a common narrative trope that organizes conflict around a metaphor of disability but ultimately evacuates the lived problem of impairment through a cure that doubles as a social fix.[42] This social fix literally requires the death of the trainer, the ape, and the doctor in order to resolve the violence of transferring vital potential from some bodies to others through immune engineering. The appropriation of national immunity in the face of the paralysis of white femininity notably appears in film at a time when nuclear fears increasingly brought about the specter of total annihilation. The film seeks resolution between medical promise and medical vampirism, and can only do so by visually displaying the instrumentality of the government of species, which penetrates deep bodily structures of both the privileged patient and the vivisected laboratory subject.

Becoming Alien: The Gaze of the Rhesus in an Era of Decolonization

At the end of the war, a series of Hollywood films set outside the continental borders associated apes and new technologies with the racial crucibles of the colonial jungle. Bioprospecting white protagonists toured jungle habitats to collect the newest cures and technical advancements of nuclear age, as in *White Pongo* (1945) and *Jungle Jim* (1948) as well as the serial *Jungle Queen* (1945). Especially in the prevalent associations of nuclear research with the horrors of the colonial jungle, these films suggested anxieties over the possible threats to security posed by emerging technology. However, with a new American exceptionalist enthusiasm for technology following the nuclear genocide in Japan and the successes of penicillin and polio vaccine, the laboratory soon became a site of comedy, realizing the transference of colonial sensationalism into domestic humor anticipated in the *Life* editors' caption of Mieth's *The Misogynist*. Far afield from the fears of polio and human-animal hybridity in *The Ape*, the hit comedy *Monkey Business* (1951), starring Cary Grant, Ginger Rogers, and Marilyn Monroe, and *Bedtime for Bonzo* (1952), starring Ronald Reagan, move into a decidedly modern laboratory, celebrating the intentions (if not the abilities) of the white male scientist rather than depicting the lab as a space of vivisection and vampirism. In this celebratory take on US imperial technoscience, apes are domesticated as children, kin within the nuclear family. *Bedtime for Bonzo* provides an apt idealization of Harlow's humanization of the laboratory primate, as the chimpanzee is a psychology professor's model for an experiment to prove that the nuclear family can instill moral values.

Monkey Business, which centers on a domesticated chimpanzee, demonstrates that Hollywood paid little attention to minding taxonomic distinctions between monkeys and apes; yet NIH officials, concerned with the connection of research animals with the lengthy histories of racialized primate representation, began actively promoting the broader designation "primate," a term inclusive of humans, apes, and monkeys. Keeping apes (primarily chimpanzees) in the Regional Primate Centers run by the NIH served to market the project as future oriented, sidestepping racialized associations of Asian and African monkeys with mimicry and aggression, sideshow performance and the colonial jungle. In a 1957 meeting of the NIH primate committee, one expert explicitly addresses the politics of taxonomic language. He argues for considering including "anthropoid apes" in the agenda for the primate centers primarily because of the "public relations" advantage of not being associated only with monkeys: "There are certain advantages of considering them [apes] in this picture, if for no other reason so that they won't be labeled as the monkeys. It helps to support the word primate. It has a lot better public relations value."[43] These statements reveal the constructedness of the category "primate" in national policy on biomedical research. To advertise "primate research" rather than rhesus research, monkey research, or chimp research avoids a range of racialized associations with Asia and Africa.

Yet just as the NIH was adopting this bureaucratic language, another unexpected outcome of the primate importation schemes was taking shape in Puerto Rico: the escape of captive monkeys and their resulting public transformation into a symbol of occupation there. The acts of monkeys themselves complicated the efforts of scientists and filmmakers to contain the figure of the monkey to a domesticated, captive body primed for both affection and vivisection. Rhesus and patas monkeys imported to Puerto Rico for scientific use escaped their semi-free-ranging colonies, remaking ecological and social space.[44] As in India, the free-ranging monkey came to be viewed by many Puerto Ricans as a pest, signaling the ambivalent imperial legacies of primate research institutions on the archipelago.

In addition to the colony at Cayo Santiago, research grants allowed Puerto Rican scientists to expand free-ranging primate colonies after the polio scare. In 1962–63, the School of Medicine's Laboratory of Perinatal Psychology, which operated as a captive colony of Cayo Santiago–derived rhesus in San Juan, established a free-ranging colony on the islets of Cueva and Guayacán, just off the coast from the southwestern town of La Parguera. Because of the extremely close proximity of these islets to the main island—shores were as close as 50 meters—escapes occurred from the beginning of the program. Escapes of rhesus and African patas monkeys, which were introduced at La Parguera in 1971, steadily increased free-ranging monkey populations in the Lajas Valley of southwestern Puerto Rico. When the Food and Drug Administration (FDA) became interested in La Parguera, the situation got worse. The FDA's 1974 grant helped maintain Puerto Rican primate facilities facing continuing funding problems. Yet with a grant to expand La Parguera, the FDA attempted to increase the colony size to 2,000 animals. Escapes increased, with entire troops of animals swimming freely across the channels of the bay. Scientists at La Parguera did not keep estimates of the number of escapes, but at the end of the 1980s, they had trapped over 250 monkeys in the southwest of the main island and reported other monkeys living free in cattle-grazing areas on the

Sierra Bermeja mountain range. The total population of free-ranging monkeys is today likely over 1,000.[45]

Patas and rhesus monkeys began to serve as figures of an invasion by the late 1990s. In 1998, news reports claimed that farmers faced 20 percent losses and were switching from profitable fruit and vegetable exports to less profitable crops and, in some instances, leaving the business altogether.[46] The new media attention was not without its exaggerations. The numbers of monkeys were regularly reported to be significantly larger than the population surveys report, and a number of economic factors—most notably the pursuit of neoliberal trade policies—placed increasing pressure on crop producers to convert to monoculture production. The monkey problem was occurring within a larger context of economic decline and a tense situation regarding the presence of the US military. It is within this context that the most sensational stories of Puerto Rico in the 1990s were disseminated internationally. The legend of the cryptid *el chupacabras*—first reported in 1994 in Canóvanas, northeastern Puerto Rico—quickly spread to the southwest, a region that in the 1990s became the epicenter for reports of paranormal activity on the archipelago: UFOs, alien landings, abductions, and the deaths of farmed animals associated with precision bloodletting. In this context, escaped monkeys were one of several signs of anxiety over the imperial presence of the US occupation; while monkeys appear to be incorporated into the visual form of this cryptid, they were also consistently presented as the real beings behind the scare, mobilized to dismiss the supposed superstition of rural Puerto Ricans.[47] This was common even as drought and other economic-environmental factors threatened farmed animals. For example, in response to a report of two sheep deaths in Lajas in 1996, officials quickly claimed that the animals had been attacked by monkeys in the area.[48]

Imagined as an extraterrestrial vampire that blends characteristics of reptiles, dogs, and primates, the chupacabras legend spread first to Mexico and then throughout the Spanish-speaking communities of the Americas; it became a staple of US televisual representations of the paranormal during this time. According to Lauren Derby, the chupacabras legend must be understood as part of a broad "culture of suspicion" regarding the US presence in Puerto Rico.[49] Derby mentions in particular the pervasive and secretive US military installations as promoting a particular "state effect": "The state in Puerto Rico is . . . pervasive yet remote; commanding yet invisible, since much of the actual muscle of US imperial power resides on the island because the

US armed forces have enormous holdings on Puerto Rican soil."[50] Economic and environmental concerns went hand in hand with suspicion over the military presence. Chupacabras sightings clustered in areas associated with US government or industrial presence, including new post-NAFTA pharmaceutical plants that sprang up in the 1990s. In the agricultural southwest, facing a prolonged drought, the phenomenon was linked to paranormal activity at the site of a new US military radar project used for the monitoring of drug trafficking. The United Front for the Defense of the Lajas Valley formed in part to oppose the siting of the US military radar project in Lajas.[51] Economic, environmental, and military forces combined to promote a civic skepticism whereby certain open secrets of the government formed an alien spectacle. Thus, chupacabras sightings in the southwest cannot simply be dismissed as superstition. As Robert Michael Jordan notes, key socioeconomic forces gave rise to legend of el chupacabras, first in Puerto Rico, and then elsewhere in the Americas: "perceptions of US economic, cultural, and political imperialism," "pollution," and "fragmentation of rural society" wrought by the post-NAFTA spread of industry to rural areas.[52] Although within a decade interest in the chupacabras in Puerto Rico had mainly become limited to paranormal and cryptozoological communities, the legend's emergence demonstrates the ways in which nonhuman animals become incorporated into complex affective negotiations over colonized space.

As farmers' protests grew louder and as public health officials raised concerns about contagious disease risk, the government took increasingly strong measures to control and eliminate monkey populations. In 1999, a new wildlife plan established the authority to manage invasive species through a variety of nonlethal and lethal means, including prescribed hunting. While trapping was ongoing in Lajas to remove animals during the last decade, there was no coordinated study of it until 2006, when the government first proposed significant funding ($1.8 million) for monkey removal. In 2008, Puerto Rico gained international media attention when it initiated a trap-for-export program, beginning with the transfer of fifteen rhesus to a private safari park in Florida; when the animals escaped the park by swimming across a moat, they led county officials on a six-month chase. No other institutions were willing to take more trapped animals until early 2009, when Iraq's National Zoo in Baghdad agreed to take a shipment of monkeys for public display. International animal rights groups denounced the measure for placing rhesus in a war zone. By December 2008, the government was openly shooting trapped monkeys to prevent their spread across Puerto Rico.[53] The monkey control initiatives of the DRNA (Puerto Rico's Department of Natural Resources and Environment) signal a reversal in the image of the monkey as an indicator of universal scientific progress. Dispersed geographically through an unequal imperial history of medical research, rhesus are alternately figured as kin and pests.

In the midcentury transformations of disease control, pharmaceutical economies, and medical science, both poliovirus and rhesus macaque bodies played central roles in the mediation of space, embodiment, and epidemic disease. If in the midst of the polio scare, the pharmaceutical appropriation of nonhuman primates energized unpredictable public fears of medical experimentation and species hybridization, these primates were nonetheless domesticated and materially integrated into national research infrastructures. The global push for decolonization set limits on national appropriation of primate bodies in locations as far afield as central Africa, India, and Puerto Rico. Because this required researchers to take control of reproduction, a laboratory apparatus that could domesticate these creatures was an important component in the public reframing of primate bodies and extractions as domesticated animal capital. As biomedical and filmic representations turned rhesus from exotic, racialized objects into docile national subjects, rhesus played integral roles in engineering the embodied form of a settler society that accepted increasingly radical biomedical intervention into patient bodies through marketing of pharmaceuticals even as it came to see humans as kin of other primates. The conscription of rhesus as defenders of the nation against dread diseases, however, produced conflicted entanglements of humans and monkeys in Puerto Rico that signaled the ongoing imbrication of primate research in the neoliberal settler politics of security. As such, the continuing migrations of rhesus macaques across US imperial borders—as far afield as rural Puerto Rico and Iraq—point to submerged transcolonial histories of the government of species that have been largely foreclosed in American exceptionalist narratives of medical progress in the antibiotic revolution. It is from this perspective that we can understand how rhesus macaques, as bodies deployed on the front lines of neoliberal disease control, have been alternatively figured as vivisected beasts, feeling kin, and alien vampires traversing the borders of empire.

Coda: Captivity and the Nonhuman Primate

I end this chapter with one final comment on the legacies of primate importation for the government of species. The transcolonial history of the NIH primate centers helped to produce an imperialist association of primates, human and nonhuman, as interspecies kin. This had an effect on languages and tropes that today guide animal activists and animal studies scholars. In one rough draft of Tulane University heart researcher George Burch's first proposal for the national primate centers, a key word is crossed out by hand in the first sentence: "sub-human." Burch handwrote "nonhuman" in the space above to describe primates in more neutral terms.[54] The term "sub-human"—used widely by Harlow and many other primate researchers—was present in many of the early documents associated with the founding of the NIH primate centers, but was eventually replaced in bureaucratic language with "nonhuman primate." As explained earlier, the primate committee was deeply concerned with public relations, which led them to use the term "primate" rather than "monkey." The bureaucratic domestication of laboratory primates as nonhumans who share an interspecies set of physical and intellectual capacities with other primates including humans was publicized in order to serve a particular set of institutional goals. These goals were less concerned with broadly questioning an anthropocentric hierarchy of species, and more involved with justifying vivisection on a mass scale. Four million animals were used in US research annually at the time of the committee's deliberations.

A properly Darwinian adjective, "nonhuman"—which today modifies "animal" in the writings of animal activists, animal rights lawyers, and animal studies scholars—gained currency within the security state out of a set of needs to institutionalize primate research, which required recognizing the affective and biological labor that made rhesus appropriate models of the human. The fact that the government of species set the stage for this phrasing used in present-day language of animal rights and animal studies does not invalidate animal advocates' ethical gesture of including humans within the sphere of animality. However, the historical situation in which claims of primate likeness are used simultaneously to justify vivisection and to oppose it emerges from the particular articulation of dread life I have described in this chapter, wherein white settler anxieties over technologies and over contact with the third world required a linguistic and visual domestication of different species of primates used in labs.[55] This colonial dynamic had unintended consequences of helping to articulate languages of animal rights and liberation that would later question species divisions once apes and monkeys could be refigured as "kin of humanity."[56] Lisa Lowe has argued that for colonial liberalism, "the human" describes not a universal species but a linguistic formalism that abstracts dynamic intimacies of labor crossing continents into a securitized, propertied form of subjectivity.[57] The rise and dissemination of the bureaucratic formalism "nonhuman" may perform a similar abstracting function, allowing the security state to rearticulate the spatial intimacies and violences of the colonial primate trade as an ideal and universalized form of neoliberal kinship and immune conscription.

Since the 1970s, research institutions have responded to animal liberation activism (including a number of incidents in which activists released monkeys and apes from research labs) with a militarized securitization of laboratory space, a locking down of those domesticated macaque and ape conscripts who had been imported and nationalized as almost, but not quite, human. This shift was intensified by the post-9/11 passage of the Animal Enterprise Terrorism Act in the United States, which cynically criminalizes public speech against vivisection and other violent treatment of animals.[58] Even as animal activists adopt tropes of human-animal likeness and kinship that were promoted by the security state during the Cold War, their use of these tropes to contest the unrestricted transformation of animals into biocapital draws the surveillance, censorship, and violence of a state that folds animal research into an increasingly aggressive agenda for biosecurity. This confrontation was anticipated in the very logics of the turn to primate modeling in mid-twentieth-century medical sciences that sought to contain the vulnerability of the national immune system in a world of proliferating war and technological competition. At that time, imperial medical science looked into the eyes of the monkey to find a deep evolutionary kinship of feeling and mobility, a shared affective space for human and monkey nervous systems to interface; it then stabbed the monkey in the back, extracting its immune capacity for a new biopolitical formation of empire.

CHAPTER 4

Staging Smallpox

Reanimating Variola in the Iraq War

The world and all its peoples have won freedom from smallpox, which was a most devastating disease sweeping in epidemic form through many countries since earliest time, leaving death, blindness and disfigurement in its wake and which only a decade ago was rampant in Africa, Asia and South America.
—Resolution on the eradication of smallpox, Thirty-Third World
Health Assembly, 1980

On May 8, 1980, the World Health Organization (WHO) passed an unprecedented resolution. Following the so-called antibiotic revolution, classic public health campaigns had finally won the battle to exterminate one of the world's most persistent and feared diseases: smallpox. A surprise even to public health officials, smallpox (*Variola major* and *minor*) had within the span of a decade become the first microbial species to be completely exterminated in the environment, unlike other long-standing targets of eradication such as yellow fever and malaria. In WHO's triumphant rhetoric, a species became the object of a declaration of independence, a proclamation that the mythic human, totalized in the diversity of its peoples, could be free from a debilitating and deadly viral pathogen. This meant the end of a disease represented as "sweeping in epidemic form" and that had long appeared to bring death and debility at jarring speed.

Smallpox has not reappeared in the environment in the intervening decades. Nonetheless, as drug-resistant forms of old diseases and the emergence of new, deadly viruses associated with neoliberal globalization tempered the postwar euphoria of disease control, health and defense officials in the United States became preoccupied by the fear that smallpox could escape the lab and become a global threat through either accidental release or intentional use as a bioweapon. As explained in the introduction, Donald Rumsfeld publicly promoted voluntary vaccination in the US media in the lead-up to the 2003 Iraq invasion, attempting to bring home to the site of the skin the virtual vulnerability of smallpox's epidemic threat. Despite its status as extinct, smallpox oddly became the top concern of US biosecurity planners at the outset of the new millennium.

In this chapter, I analyze the fantasized reemergence of epidemic smallpox as an incitement to war in the late twentieth and twenty-first centuries. Smallpox, I argue, accrues its power in part through its status as dead in the environment, and is thus imaginable as "sweeping" undeterred through vulnerable, nonimmune populations. Like Hansen's disease as described in chapter 1, the sensational state view of variola attributes to the virus the zombie status of the animated dead. It produces a "leaky," hemorrhaging body, sensationally depicted as a prone, blistered, and feverish product of human-viral hybridization, requiring emergency containment. As pop-science writer Richard Preston writes of variola's most deadly form, hemorrhagic smallpox, "The skin ... could slough off in sheets when nurses tried to move a patient. The patients had hemorrhages from the mouth and nose, intestinal tract, and urinary tract. The whites of the eyes of a hemorrhagic smallpox victim would turn bright red from leaking into the eyes. Smallpox was a monster."[1] Still worse, in the paranoid visions of defense planners, smallpox completes a quantum leap from its history as a colonial bioweapon to a future in which it rapidly sweeps away life along the viral routes of globalization, spurring terror in its wake. This vision of apocalypse is deeply connected to anxieties about the changing territorializations of US empire, which neoconservatives have proposed must engage in total war to intervene in inner and outer spaces, a full-spectrum dominance in defense of a mythic Pax Americana. This figuration pits hope in technology against the viral devolution of the settler homeland, imagined through the stereotyped figure of a vanishing American Indian.

Saddam Hussein's suspected weaponization of smallpox was, notably, one of the main justifications for the 2003 US occupation of Iraq. While the Left in the United States and Europe has understandably dismissed such incorrect US claims about Iraqi weapons as cynical lies perpetuated for unrelated political, economic, or strategic interests, this line of Left criticism risks sidelining attention to some of the broader mutations in the connections between race, contagion, state secrecy, and imperial security that stubbornly tie both mainstream political parties in the United States to forms of interventionist idealism. The Left's explanatory framework often instead shares with the Right a familiar model of colonial antagonism: a Newtonian rule of force in which every action produces an opposite reaction, whether containment (the Right) or blowback (the Left).[2] The present chapter departs from this formulation, articulating a queer critique of neoconservative war doctrine by exploring its reproduction of forms of racial suspicion and fears of bodily transition that circulate between publics and institutions such as media and the security state. Analyzing public figurations of sweeping bodily vulnerability, racialized contagion, and imperial speculation related to smallpox in the construction of the Iraqi bioweapons threat, I work to understand how the government of species catalyzes a politics of scale in addition to its earlier emphases on space and time. Variola, an alien species that represents the virtual extent of human biological vulnerability, sweeps through bodies and figures—including the laboratory monkey, the vanishing Indian, and the Islamic rogue terrorist—in contemporary debates over biodefense, scaling up from contained microscopic threat to reveal the global political and biological uncertainties of a new post-Soviet era of empire. These potential viral contact points intensify the affective potency of smallpox, helping state planners envision its stickiness to populations and the threat of its reanimation.

This dynamic is intensified as figures of viral threat circulate across the public-secret informational divide managed by the security state. Whereas the Edward Snowden affair and the disclosures of Bradley Manning and Wiki-Leaks have recently commanded much journalistic attention concerning the limits of public disclosure and the forms of secrecy that appear to be ever expanding in an era of digital warfare, the Iraq bioweapons debacle suggests that the liminal category of "sensitive but unclassified" information (including such information on smallpox and other suspected biothreats) may be an equally potent site for the articulation of neoconservative warfare through information economies. My analysis thus begins in the worlds of military biosecurity, of secret biotech labs populated by microbiologists, machines, monkeys, and microbes; it then fans out into literature and journalism, where a small cadre of writers worked with a group of biodefense officials intent on publicizing the threat of Iraqi weapons proliferation in the 1990s and 2000s. I call these writers "literary leakers" because they translated sources' partial summaries of classified intelligence into relatable and legally disclosable public speculations on bioweapons risk. This effectively transformed uncertainties in intelligence into a form of suspense that drives the speculative form of literary narrative. Although their speculations proved to be completely divorced from the reality in Iraq, I explain how literary leakers created fiction and journalism that became actual weapons of the security state.

In the aftermath of the official occupation, with no confirmed unconventional weapons in Iraq, the neoconservatives claimed that the very absence of Iraqi biological, chemical, or nuclear weapons demonstrated the prudence of the policy: the United States acted before an imminent threat could materialize. According to what Arjun Appadurai calls a "vivisectionist" logic of war that attempts to produce knowledge through violence, Iraq no longer needed to be the site of the "nightmarish task of divination" of proliferating small-scale biothreats.[3] Yet the Bush doctrine of preemption was not in itself enough to catalyze the intensity and momentum of support necessary among different state actors and media to mobilize invasion and occupation. Other bodies—bodies that could transit through and amplify Saddam Hussein's imagined force and geographic reach—were necessary to catalyze action; war had to be translatable across interest groups and appeal to a broad sense of insecurity. Racial fear, translated as the vulnerability of humanity to a virtual, alien species, was deployed to help intensify the public sense of insecurity. Variola today exists only in freezer safes at government-run laboratories in the United States and Russia. Yet in the speculations of journalists and defense officials, it transits through various systems, spaces, and times, emerging in fantastic journeys across the paranoid routes of imperial enemy mapping, intensified by the stratified practices of state secrecy within the military and political apparatuses. Variola's very eradication signaled its potency, its unpredictability—features that lined up with racial fears of Saddam Hussein's potency, his uncontainability within national borders, and his unpredictability as an isolated rogue leader.[4] It is intrinsic to neoconservative logic of governmental complexity that action will produce its own instability, that there will be some blowback to every action, but not in Newtonian form; actants jump across space and time in viral leaps of scale, creating new forms of disorder in which the US imperial state invests hope due to its privileged capacity to globally respond, to always have the resources for more redirection if never complete containment. Controlled disorder produces internal limits to the practice of full-spectrum defense, as "dual use" of bioweapons research and the risks of prophylactic measures like vaccines have threatened US biosecurity initiatives from within. Despite this, the efforts to combat smallpox and other biothreats have exposed the expanding sphere of potential vulnerabilities that the security state attempts to survey and contain across species, systems, environments, and borders. This suggests the ongoing blurring of conventional, bio-logical, environmental, and informational warfare in the twenty-first century.

Smallpox Emergence

The emergence of variola as a perceived risk to the state and the species is deeply immersed in neoliberal forms of thinking about disease. As explained in the introduction, US global health specialists coined the emerging diseases paradigm to signal heightened awareness of transborder disease risk coincid-ing with the economic and environmental transformations of post-Fordist globalization, which I explore here and in chapter 5. In the aftermath of the rise of emerging diseases, explains Andrew Lakoff, preparedness strategies in-vite heightened forms of speculation and surveillance to the calculus of bio-logical threats.[5] In the militarized language that portrays redoubling microbial enemies jumping the temporal and spatial scales of globalization, emerging diseases are constructed as the near future's biological threats to human popu-lations that cross borders with increasing speed and frequency. Expanding on earlier imperialist articulations of eradication and prophylaxis, the emerg-ing diseases model seeks dominance of particular environments and trans-national conditions that give rise to the spread of disease. At the same time, disease control is militarized not only through technical language but also through the actual incorporation of disease policy into military planning at the highest levels, exceeding the interpenetration of medical and military apparatuses that existed a century earlier in the heyday of colonial tropical medicine. As such, state approaches to disease have increasingly abandoned traditional cost-benefit analyses that public health officials used to triage the most probable threats to populations. It is no longer possible to be prepared for the next catastrophic disease (like Ebola or AIDS) by using existing knowl-edge; the public health infrastructure and other vital systems have to be ready to generate emergency knowledge on and responses to the next big bug, the coming generic threat on humanity's horizon.

In her seminal essay describing new biosecurity paradigms, Melinda Cooper suggests that "the biological turn of the war on terror" consists in a broad array of state-based attempts to establish preparedness not against par-ticular disease threats, but rather against the process of *"emergence itself."* For Cooper, this militarized approach to microbial pathogens "defines infectious disease as *emerging* and *emergent*—not incidentally, but *in essence."* This approach to disease makes it one domain in a broader neoconservative attempt at full-spectrum defense, an agenda that suggests "the scope of security should be extended beyond the conventional military sphere to include *life itself* . . . from the micro- to the eco-systemic level."[6] In 2000, the CIA declared that disease posed a threat on par with terrorism, and President Clinton declared AIDS a national security threat.[7] In 2002, George W. Bush made the connection between smallpox and terror explicit: "To protect our citizens in the aftermath of September 11, we are evaluating old threats in a new light. Our government has no information that a smallpox attack is imminent. Yet it is prudent to prepare for the possibility that terrorists . . . who kill indiscriminately would use diseases as a weapon."[8] This assessment was confirmed by a bipartisan panel on weapons of mass destruction (WMD), which claimed in 2008 that bioterrorism was "more likely" than nuclear terrorism and thus should be made "a higher priority."[9] While the emerging diseases paradigm shared with earlier colonial disease eradication and prevention efforts a focus on territorial risk, a reliance on biomedical commodities for prevention, and sensational images of disease invasion from Africa and Asia, this paradigm made the speculative identification of threat areas and systemic vulnerabilities its dominant targets of intervention; future risks were straightforwardly functions of the scope of present action.[10] Preparedness in this scenario stages catastrophic risk as an everyday phenomenon, fully integrated into the normative functions of the security state even as neoliberal austerity means that the state retrenches other areas of governance.

Yet WHO has declared smallpox eradicated, extinct in the environment. So in what sense is smallpox also emerging? Variola's environmental absence means that individual human immune systems may lose their acquired "memory" of variola developed during the era of mass vaccination, which ended for the United States in the late 1970s. Because smallpox is a highly contagious virus, this reliance on distributed immunity makes a smallpox epidemic appear to be a more plausible scenario of attack than other potential bioagents. Consider the contrast between smallpox and anthrax (*Bacillus anthracis*), which is considered the most likely biological weapon due to its wide availability and ease of deployment. An anthrax attack circulated in the US mail immediately following 9/11, most likely from a research scientist who had the proper security clearances to access the US government strain that was mailed to the offices of journalists and politicians. Anthrax works within a standard territorial logic of weaponry since it can be deployed in a limited geographic region against an enemy.[11] Yet because of anthrax bacteria's vulnerability to exposure, it is unlikely to reinfect a large second generation of victims either through human-to-human transmission or through zoonosis between humans and bovines. *Variola major*, conversely, could theoretically transit beyond the index case and move across networked human populations through airborne transit.

In the scenarios of neoconservative planners, smallpox has helped stitch a logic of defense from nuclear containment during the Cold War to the fantasy of full-spectrum dominance during the war on terror. The neoliberal model of disease prevention associated with polio in the postwar period (chapter 3) generally placed the onus of protection on a biomedical consumer whose value-added immune system was optimized by a national research apparatus that invigorated a private market in pharmaceuticals. Today, the neoconservative model suggests that the state must be able to deploy rapidly to anticipate and engineer high-tech solutions to the most catastrophic threats, marshaling emergency research and generic capacity to mass-produce vaccines, publicly advertising the power and necessity of quarantine and martial law as paths of last resort. Furthermore, in an era of fiscal austerity, this approach means that the security state takes on an outsize share of public resources, coming into budgetary conflict with more traditional public health and social welfare institutions. Producing urgency of militarized defensive response against apparently catastrophic agents justifies ongoing reform of and inflationary investment in the security state as if the uncertainty inherent to life was itself the enemy. Bypassing cost-benefit risk calculations and broad social approaches to health, preparedness means targeting complex speculations of catastrophic harm, such as the unlikely scenario that a terrorist organization or rogue state would acquire extant virus stocks, research apparatus, and delivery technologies required to deploy weaponized smallpox.

Playing Indian: Smallpox and the Suspended Animation of American Empire

Settler colonial racialization plays a particular role in the publicity of smallpox's catastrophic risk. State speculation on the reemergence of smallpox since the eradication has transited through at least two distinct but interrelated figures of racial intensification: the vanishing American Indian and the Islamic rogue terrorist. To understand the connection of smallpox to these racial tropes, we will have to begin with the microworlds in which viruses do their work on animal bodies. Viruses are biological particles that exchange

DNA across larger bodies, playing potentially important roles in the mutations that occur in ecological adaptation. They become highly adapted to groups of hosts—particular bodies in particular ecologies—and replicate by jettisoning their genetic matter into living cells, which fuse into the genetic structure of those cells or otherwise generate proteins and genetic code that may facilitate viral proliferation. Microbiology has historically understood this relation to the cell's biosystems as a form of parasitism.

Another way to understand this relation, however, is that viruses work ecologically on the plasticity of bodies, introducing play with the limits and potentials of the bodily forms of hosts. Poxviruses—which infect a wide variety of animal species from cows to monkeys to mollusks—offer an apt illustration of bodily plasticity. Transitioning into a variety of forms as they cross into and inhabit different bodies and species across evolutionary time, poxviruses may have little impact on some hosts but stress the organs, borders, and solid structures of other bodies. In caterpillars and other insects, poxvirus produces systemic distention of the body up to twice its size, literally demonstrating the liquidity of internal organs and skin. In humans, smallpox (which is closely related to cowpox and likely evolved through human-bovine domestication) is known for its production of disfiguring rashes and blisters across the body, covering limbs and intensifying in number around orifices such as the mouth and the rectum. With the accumulation of virus, inflammatory responses shift into overdrive. In its hemorrhagic form smallpox also produces mass internal bleeding (forging black patches of bloody skin) and massive hemorrhages in the organs. Stories of the disease thus envision a leaky human, who experiences severe fever, inadequate clotting, and expulsion of blood and black pus. The plastic play of smallpox in relation to the human does not produce uniform results—weaker strains of the virus tend to produce less severe bodily impacts, and varied formations of immunity among individuals and groups modulate viral play with bodily structures. But smallpox produces a type of pressure and liquidity that pushes bodily structures to their breaking points—in some instances bringing about death, in others producing a flurry of painful and debilitating symptoms before immune responses moderate viral expansion.

Historically, this ecological variation in the embodied play of smallpox has both produced racial knowledge and lent power to the use of smallpox as a weapon. European colonists in North America viewed the contrast between a growing settler population and an immunologically naive native population as a racial sign of divine right to land. In the late nineteenth century, Chinese immigrant communities became targets of rhetoric linking smallpox to racial risk.[12] Smallpox is one of the most deadly diseases in human history, but its variation in populations means it unevenly impacts individuals and becomes a site at which social differences such as those of race and nation become objects of surveillance. The WHO eradication declaration quoted in the epigraph to this chapter hints at the horror of smallpox epidemics, noting the disease's "sweeping" spread through "many countries since earliest time." In the process, WHO attributes sweeping agency to the virus itself and collapses the divergent ecological conditions of colonial settlement whereby smallpox was made into a genocidal agent, the world's first large-scale bioweapon.

Eradication has done something unique to the affective context of imagined human-variola contact: since smallpox exists not in the queer movement and interaction of life, but in the suspended animation of liquid nitrogen frozen in state laboratories, it no longer coevolves with humans. The smallpox we would encounter were it to be weaponized would be archaic, literally alien to our world given the loss of decades of adaptation to its environment in the human body. Researchers can only speculate on the speed in the decline of immunity and the differences in the potency of different natural or engineered strains of variola if they were to emerge. Since eradication has allowed states and global health entities to stop vaccinating for smallpox, public health officials sometimes assume that immunity in the United States and elsewhere has declined to levels prior to the settler colonization of the Americas (a shaky assumption given the centuries of human coevolution with the virus in its endemic regions). They further assume that with intensified economic integration, travel, and migration, smallpox would quickly be able to gain a strong ecological niche of the sort that allowed it to spread in epidemic form in prior centuries. If the world has stopped vaccinating against smallpox, its infectious potential begins to appear more like the smallpox threat that famously faced the American Indian, who, according to the present outbreak narratives, lived in a viral *terra nullius* when Spanish and, later, French and British settlers arrived to occupy Indian nations.

Smallpox played a role in the genocide of Indians in Mexico, Central America, the Andes, and the Caribbean, often moving just ahead of the supposedly heroic conquistador. The unfamiliar virus—in the absence of indigenous immunity that would come with regular contact with humans or bovines from endemic regions—likely infected large numbers of people across the Arawak, Aztec, Mayan, and Incan lands. In the process, variola heightened the impact of other infectious diseases, challenged military defenses, and produced visual spectacles of blistered and bleeding bodies that threatened existing social organizations.[13] In the years preceding Edward Jenner's experiments on children to prove the efficacy of smallpox vaccination in 1796, British North American settlers benefited from smallpox epidemics, including those spurred by the intentional infection of Indians. Especially in the two decades leading up to the Revolutionary War, smallpox epidemics moved through the Carolinas, the High Plains, and the Pacific Northwest, unequally impacting Indian nations; meanwhile, documentary evidence of the French and Indian War indicates that Sir Jeffrey Amherst and Colonel Henry Bouquet connived to spread smallpox among thousands of Ottawa and other Indians of the northeast as one of the world's first intentional bioweapons.[14]

Public health historians often collapse such histories of the uneven territorialization of smallpox, engaging in a form of universalist species thinking about human vulnerability common to post-1970 global health paradigms: "Smallpox was a democratic scourge, afflicting people of every race, class, and social position. The disease killed royalty as well as commoners . . . and repeatedly [changed] the course of human history."[15] State and academic accounts conjure the Indian as a paradigmatic example of today's global decline in immunity. In the volume *Emerging Viruses*, adapted from papers of the field-defining 1989 conference led by microbiologist and biodefense planner Joshua Lederberg, medical historian William McNeill situates the bodies of settler-colonial Americans in the position of biologically "virgin" Indians: "I had enough general understanding of what an infection such as smallpox let loose on a virgin population might do to be able to imagine what was going on in Tenochtitlan that night. It wasn't only Montezuma who died. Aztec civilization died too. . . . Indeed, the whole history of our country and of the Americas at large . . . is a function of disease disequilibrium that existed after 1500 between the Old and the New World."[16]

In the consensus statement of smallpox eradicator D. A. Henderson and other US biosecurity officials in their 1999 review of the smallpox weaponization threat, the nonimmune Indian provides the sole historical example of the potential future effects of epidemic smallpox; the weaponized use of smallpox in the French and Indian War marks a period of nonimmunity just before Jenner's vaccine mitigated the global threat. Eliding in the passive voice European settlers' responsibility for the genocidal spread of diseases in the colonial formation of the US state, the authors make the Indian body the paradigmatic natural victim of disease: "Smallpox probably was first used as a biological weapon during the French and Indian Wars (1754–1767) by British forces in

North America. . . . Epidemics occurred, killing more than 50% of many affected tribes."[17] Elsewhere, Henderson notes that when colonial-era North American cities were too small to sustain smallpox transmission, years of absence rendered areas fertile for new epidemics: "The longer the period of freedom from smallpox, the larger the number of vulnerable people and the more disastrous the epidemic."[18] Suggesting that the end of vaccination after 1977 could return humanity to the state of the nonimmune Indian, Henderson and colleagues support the controversial US stance in support of retaining the remaining laboratory smallpox stocks for defensive biosecurity research. In this rhetorical move that Jodi A. Byrd terms "the transit of empire," the figure of the always-already-dying Indian is remade into the transcontinental future of a globally expanding American empire, offering affective scaffolding for the ongoing reproduction of militarized settler sovereignty and its increasingly planetary targeting of land and life.[19] The Indian, conventionally figured as a fleeting, nonsovereign specter abandoning the terra nullius of settler sovereignty, transits to render vulnerable the expansionist body of empire now seen as threatened by emerging diseases.

Engineered Uncertainty: Bioweapons and Iraq

This racial transit of smallpox coincides with its shift from a natural to an engineered threat. Smallpox's potency no longer lies in biological reservoirs, but instead in laboratory cultures. Immediately after eradication in 1978, Janet Parker, a British medical photographer, died from smallpox traced to an unsecured laboratory at the University of Birmingham. In response, D. A. Henderson and WHO called for laboratories worldwide to cede remaining stocks of the virus to the two major repositories of the eradication, the Centers for Disease Control (CDC) in Atlanta and a Soviet facility located in Koltsovo, Siberia. Frozen in boxes within liquid nitrogen–filled safes, extant smallpox strains collected during the eradication are maintained at secret locations under the highest biosafety standards (BSL-4). Other institutions are prohibited from maintaining materials that include more than 10 percent of variola's DNA, allaying fears of the future potential of biosynthesis technologies to artificially produce an entire smallpox organism from DNA sequences. In 1986, WHO called for concluding the eradication with the destruction of the two repositories, allowing only for the storage of DNA samples for future research. Yet by 1993, this agreed date for the destruction of the laboratory reservoirs of the disease would be postponed. In a climate of mutual distrust, both the

United States and Russia balked on full extermination, leading to a series of postponements of the scheduled destruction (most recently in 2014).

The change in policy demonstrates the tautology that has characterized the posteradication smallpox threat: the risky existence of laboratory smallpox justifies its continued preservation. In the face of laboratory stocks not controlled by the United States, the state finds an indefinite research need to combat the uncertainty of engineered smallpox elsewhere. Such a circular logic was already in place by the time in 1992 when Dr. Kanatjan Alibekov, former deputy chief of research and production for the Soviet Biopreparat, left Russia for the United States. Taking the Anglicized name Ken Alibek, he accused the Soviet and Russian states of offensive bioweapons research, carried out partly under his own tenure. Moscow had made secret stores of anthrax and had experimented with recombinant DNA research to increase variola's resistance to drugs. Yet Alibek's allegations went further, claiming that Russia sought to genetically engineer a highly virulent epidemic strain, that it had developed ballistic missiles with warheads capable of disseminating smallpox and anthrax, and that Soviet scientists had shared knowledge with Iranian and Iraqi scientists. Some of Alibek's allegations, particularly concerning the existence of large biological fermenters, were contradicted by international inspections of former Soviet facilities; others were speculations on intent based loosely on research openly published by Russian scientists years after Alibek's departure.[20] These allegations nevertheless stirred old Cold War passions among US policy experts. Henderson et al.'s 1999 statement on smallpox weaponization cites Alibek's warnings as the primary justification for the US retention of smallpox stocks, although more recently journalists have increasingly called into question Alibek's motives and claims.[21]

While the Russian weaponization threat (and Russia's reverse allegations against the United States) could have been taken as incentive to destroy the remaining variola stocks, the United States and Russia instead convinced WHO to maintain the stocks, which are in practice administratively nationalized rather than international. Both countries have engaged in research that could be labeled "dual use," having both offensive and defensive applications in warfare. Russia was aware that the United States had stockpiled several biological agents during its twenty-six-year bioweapons program started by Franklin D. Roosevelt; Russian attempts to weaponize and stockpile anthrax and smallpox during the Cold War are part and parcel of this longer biological arms race. The speculative hermeneutic of risk assessment regarding state bioweapons programs means that there is no fundamental distinction between offensive and defensive research. Any defensive advantage (such as apparently peaceful vaccine research) enhances offensive capability because it decreases one nation's vulnerability to the agent. Yet the national media's presumption that US bioweapons research is always defensive in nature has minimized domestic reporting of dual-use capacities in the US biodefense infrastructure. Hawkish defense analysts, including Paul Wolfowitz (undersecretary of defense for policy, 1989–93) and his protégé I. Lewis "Scooter" Libby, were important voices in efforts by military officials to convince both the Department of Defense leadership and the CDC to expand dual-use biodefense research in the 1990s. There were at least two secret dual-use programs during the Clinton administration: Project Baccus and Clear Vision.[22] Alibek, along with retired US bioweapons engineer William Patrick III, who died in 2010, helped convince US officials that the proliferation and weaponization of anthrax and smallpox were continuing defense concerns, despite the confirmed destruction of Russian and Iraqi stockpiles.

In 2000, the CDC began to allow scientists from AMRIID, the US Army Medical Research Institute for Infectious Diseases, to secretly infect crab-eating macaques with strains from the CDC smallpox stocks for defense research. While in the 1960s the US military carried out simulated bioweapons attacks on rhesus macaques at Johnson Atoll south of Hawai'i, the more recent project to intentionally infect monkeys with smallpox violated the basic WHO rules of the eradication. Smallpox was understood to infect only humans, a guiding principle of eradication since the absence of animal reservoirs meant that the disease could not ecologically reemerge. Yet AMRIID set out to prove that smallpox could infect monkeys by injecting them with huge loads of virus directly into the bloodstream. Portraying the monkeys as sacrificially laboring in the national service, similar to dogs killed in Vietnam who had "died for their country," Richard Preston's account of this research reads like a suspense novel, sentimentally detailing—and dismissing—military researchers' ethical dilemmas over killing monkeys and intentionally developing an animal reservoir for a deadly pathogen. Preston recounts in detail the use of hazmat suits, Lysol showers, segregated oxygen supplies, intensive isolation, and tenting that prevented incidental infection of humans or monkeys with the aerosols and syringes used in the secret laboratory. Each syringe was loaded with "a billion particles" of smallpox in an attempt to bypass the initial infection and immune response stages of the disease. Skipping to the viral amplification period in humans, the direct blood-borne infection modeled the period in which the virus replicates in huge numbers before either petering out or killing the host.

Four days into the experiment, monkeys began dying, displaying what one researcher described as "spotty, starlike red spots all over their skin. . . . The animals had flat hemorrhagic smallpox. *My God, it was bloody.*" Zooming in on the "bloody" stomach, "speckled" lungs, "dead" liver, and "ultraswollen ball" of a spleen during AMRIID scientist Lisa Hensley's autopsy of one monkey, Preston takes the violent effects of the amplified virus as a sensible demonstration of the potency of variola's threat: "She was face-to-face with *variola major* for the first time in her life. Until she had seen this hemorrhagic monkey, she had no idea how powerful the virus was, how truly frightening." Of course, there was no environmental precedent for such an amplified outbreak of any disease. Justifying the introduction of smallpox into a new species by weighing "tens of monkeys" against the virtual threat to "tens of millions of humans," AMRIID chief research scientist Peter Jahrling is quoted as defending the sacrifice of the monkeys and the violation of international dual-use rules. These experiments—in which the US military violated the Biological and Chemical Weapons Convention—were presented within the secretive defense apparatus as evidence that Russian or Iraqi research could also develop zoonotic or engineered forms of smallpox rather than a gross violation of experimental ethics.[23]

Such illegal engineering of sensible evidence of the smallpox threat, worked through the bodies of monkeys, passed through the defense establishment behind the closed doors of government secrecy. The public pronouncements of politicians proximate to these state secrets are one (admittedly incomplete) window into the intensification of the fear of smallpox as integrated into a military doctrine of force for a post-Soviet world. Alibek claimed after 9/11 that he had "no doubt" that Iraq possessed biological weapons—an accusation that was echoed in civilian public health circles by Joshua Lederberg, the Nobel laureate microbiologist who led the charge against emerging diseases.[24] This assumption was common across political parties in the United States given the recent history of US sponsorship of Iraq's bioweapons program. The United States had actual knowledge of Iraq's historical biological and chemical weapons programs, in part, because US agencies helped build these institutions.[25] During the UN inspections in the 1990s, a military scientist's defection led Iraq to disclose producing anthrax, botulinum, and other bioagents at three plants. Smallpox was not among these agents, but weapons inspectors continued to suspect that Iraq had undisclosed weapons in storage and secret connections to Russian scientists. The UN Special Commission

(UNSCOM) subsequently supervised the destruction of the declared plants and their known biological stocks.

The lack of hard evidence did not contain the fear of smallpox for the neo-conservative war hawks who had long called for the overthrow of the Iraqi government. Although the original neoconservatives—Jewish liberals and socialists from New York who defected to the Republican party during the rise of the civil rights movement—initially focused on domestic political and economic issues related to affirmative action and social welfare, they later began to focus intensely on defense and foreign policy issues in their support for Israel in its 1967 colonial war against the surrounding Arab states who were also jockeying for control in Palestine. Uniting these domestic and foreign policy agendas was an emphasis on the state's inherent inability to control systemic uncertainty, particularly in their sociology of race relations emphasizing the irreconcilable experiences of Jews and blacks at home and Jews and Arabs abroad. From this perspective, the state could no longer be reliably seen as an arbiter between social groups through policies of welfare, rights, and reconciliation; security was the proper form of interventionism given the intractable and ultimately unknowable distinctions between social groups.[26] From the beginning, the conceptualization of uncertainty that motivated neoconservative policy discourses was a racialized one, intertwining the voids of otherness long sedimented in colonial discourses of antiblackness and orientalism. These racial roots of neoconservative philosophy were later rearticulated in a focus on post-Soviet uncertainty, usually connected to the specters of Islamic terrorism and rogue states. Thus the interplay of racial difference and insecurity is inherent to the logics of neoconservative doctrine.[27]

After the neocon defense hawks were marginalized with Bill Clinton's election, they joined forces with the Project for a New American Century in 1997. The project, which became the public voice of neoconservatives, disseminated a number of letters signed by key defense officials including Wolfowitz, Libby, Dick Cheney, Donald Rumsfeld, James Woolsey, and Richard Perle and complemented by hawkish intellectuals including Zalmay Khalilzad, Francis Fukuyama, Norman Podhoretz, and William Kristol. These statements advocated for the United States to protect its "principles and interests" through military, diplomatic, economic, and political means. In its statement of principles, the project articulated the preemptive logics of imperial intervention, using the very language of emergence that Lederberg deployed in global public health: "The history of the 20th century should have taught us that it is important to shape circumstances before crises emerge, and to meet threats before they become dire."[28] In its infamous report *Rebuilding America's Defenses*, which called for the forcible military removal of Saddam Hussein in 2000, the project argues that to maintain its strategic advantage of deterrence in a "unipolar" world, the United States needs to make a spectacle of its power to "fight and decisively win multiple, simultaneous major theater wars."[29] The Newtonian concept of deterrence, classically defined by Thomas Schelling as the production of a shared interest against mutual destruction by two nation-state enemies, had during the Cold War been based on the control of separate spheres by the mirrored imperial powers of the United States and the USSR. In a unipolar world, deterrence had to be purchased in multiple regional arenas; in effect, the United States had to spectacularly demonstrate not just its might but its quickness to act across the globe and into both inner and outer space in order to fully benefit from the force of dominance.[30]

While the project publicly presented this theory of deterrence in terms of rational force projection, the US Strategic Command's theory for deterring terrorists and rogues in the post-Soviet era suggested that uncertainty was central to its governing logic. Being trigger-happy, combined with mainte-nance of a large US nuclear arsenal, had continuing deterrent effects: "It hurts to portray ourselves as too fully rational and cool-headed"; it is "beneficial" if "some elements may appear to be potentially 'out of control.'"[31] The logic could be deployed against a rogue leader like Saddam Hussein as such: the sovereign must engage in "frightening, scaring the enemy . . . by giving the rogue the image of an adversary who always might just do anything, like a beast, who can go off the rails and lose his cool . . . when his vital interests are at play."[32] Uncertainty thus marks both the incitement to and the practice of security. This logic of preempting the uncertain racialized enemy with the un-certainty of imperial force projection did not require establishment of Iraq's actual pursuit of bioweapons, or of smallpox in particular; it only required that dispersed potentials for catastrophic smallpox weaponization existed in the world, and that they could network around sites of intensification: rogue states and terrorists. Intervention was understood as a kind of deterrent, but one that required constant activity, analysis, and adjustment; containment would now be an indefinite proliferation of small and large acts of war. Nor-man Podhoretz laments that the administration placed so much emphasis on WMD, when this general logic of preemption and deterrence in his eyes justi-fied the Iraq action.[33] Karl Rove sums up the logic with blunt hubris: "We're an empire now, and when we act, we create our own reality. And while you're studying that reality—judiciously, as you will—we'll act again, creating other new realities, which you can study too, and that's how things will sort out. We're history's actors . . . and you, all of you, will be left to just study what we do."[34]

It is not the threat itself of a known biological or chemical weapon that underlines the urgency of military intervention in Iraq, but the problem of uncertainty. This was a fully theorized war using uncertainty against uncertainty, attempting to redirect the queer worlds in which states, species, technologies, and identities fail to remain dependably disentangled by borders. Iraq's very submission to the UNSCOM inspections regime only increases the uncertainty associated with its weapons program:

> Even if full inspections were eventually to resume, which now seems highly unlikely, experience has shown that it is difficult if not impossible to monitor Iraq's chemical and biological weapons production. The lengthy period during which the inspectors will have been unable to enter many Iraqi facilities has made it even less likely that they will be able to uncover all of Saddam's secrets. As a result, in the not-too-distant future we will be unable to determine with any reasonable level of confidence whether Iraq does or does not possess such weapons. *Such uncertainty will, by itself, have a seriously destabilizing effect on the entire Middle East.*[35]

Failing to act preemptively means "we will face the prospect of having to confront him at some later point when the costs to us, our armed forces, and our allies will be even higher."[36] Under pressure from the neocons and with wide congressional support, the Clinton administration passed the Iraq Liberation Act, making it official US policy to remove Saddam from power.

From February to December 1998, when Clinton and the United Kingdom launched the bombing campaign that laid the groundwork for the aerial occupation and economic sanctions of Iraq, US politicians openly discussed unilateral intervention in terms similar to the 2002 debate leading up to Bush's all-out conventional war, as Iraq decided to refuse unlimited access to UN inspectors, several of whom we now know spied for the CIA. James Rubin, Madeleine Albright, John Kerry, and John McCain all publicly proclaimed the right of the United States to act militarily to oust Saddam Hussein based on his government's apparent stonewalling of the UNSCOM mission.[37] Iraq received unique defense focus after the fall of the Berlin Wall, connecting prior Cold War imaginaries of threatening enemies with the emergent new threat of the Islamic rogue terrorist. Melani McAlester describes the unique composite figure of Saddam Hussein: he was portrayed in the US media as simultaneously an irrational and excessive threat of Islamic militancy and a conventional rogue, a reincarnation of European fascism.[38] The United States had been at war with Iraq, in different forms, since the fall of the Soviet Union. If the uncontainability of Saddam's weapons and borders posed the initial public threat in 1991, this was transmuted into aerial control and economic sanctions during the Clinton era, and finally the formal military occupation in 2003.

Racial Suspicions: The Speculative Non/Fiction of Viral Outbreak

The activities of the neocons and biosecurity officials in the lead-up to the recent Iraq War of 2003 were aimed broadly at influencing public opinion, not only through publication of specific suspicions of security threats but by the management of the divide between state secrets and public information. Weapons programs are deeply implicated in the expansion of state secrecy since World War II. The United States, as the most deeply entrenched nuclear state, uses the restriction of knowledge of weapons technology to maintain a competitive advantage vis-à-vis competitors. Driven in part by dual-use WMD research, state secrecy has broadly expanded since the onset of the Cold War. An estimated 4 million Americans hold security clearances and 8 billion pages of material were classified between 1978 and 2004. If Karl Rove was blunt in his characterization of those who study US empire as inevitably falling behind the work of the security state, Peter Galison clarifies that this is in part because the state exercises a monopoly on large segments of knowledge. According to Galison, "The classified universe is, as best I can estimate, on the order of five to ten times larger than the open literature that finds its way to our libraries. Our commonsense picture may well be far too sanguine, even inverted. The closed world is not a small strongbox in the corner of our collective house of codified and stored knowledge. It is we in the open world—we who study the world lodged in our libraries, from aardvarks to zymurgy, *we* who are living in a modest information booth facing outwards, our unseeing backs to a vast and classified empire we barely know."[39]

Since 9/11, US government agencies have increasingly invoked an ambivalent category of state secret called "sensitive but unclassified" (SBU) that denominates a wide array of government projects, technical knowledges, and publicized intelligence as inappropriate for public circulation. This category of information, which becomes a privileged technology for managing the open secrets of the security state, has a complex relation to both publicity and secrecy. As Joseph Masco notes, these open secrets proliferate widely outside the nation-state, but face increased scrutiny and censorship within it, as government agencies restrict circulation domestically. At the same time, open secrets suggest the productive force of constrained or limited display—they accrue iconic power, suggesting proximity to unknown, more risky classified secrets that inflate the rumored threat of the object that is partially disclosed.[40]

Public disclosures about the apparent threat of WMD—from Iraq but also North Korea, Russia, and elsewhere—capitalized upon the structuring of this public/secret divide, especially as information deemed sensitive was selectively released to the media. Responding to outreach from the secretive defense apparatus, a small group of writers concerned with bioterrorism shared the selective SBU speculations of military researchers and neoconservative analysts publicizing the apparent smallpox and anthrax threats from the 1990s and early 2000s. These works broke down generic distinctions between fiction, journalism, and drama, and repeatedly incorporated Iraq as an imagined transit site for biological agents to circulate across borders. Speculations grounded in the probabilities of rumor and partial, unvetted knowledge appeared to pass the standards of objectivity for journalistic publications, while fictional novels incorporated apparently realistic journalistic musings on the history and science of bioweapons.

There is a racial specificity to this mobilization of the variola threat across the public/secret divide. Masco argues that racial paranoia is a structural effect of the security state's practices of secrecy. This became a problem in the failed 1999 prosecution of Taiwanese American nuclear scientist Wen Ho Lee as a suspected Chinese double agent. Despite the fact that he actually helped the FBI investigate Chinese scientists, Lee was charged and jailed over a Chinese government disclosure of US nuclear design. After his later exoneration, according to Masco, there remained a pervasive climate of racial suspicion that has continued to stymie recruiting of Asian scientists at Los Alamos.[41] Following similar Cold War logics, governmental surveillance and public reporting on proliferation threats commonly takes the foreign nationality of scientists as itself cause for suspicion, while the word of defectors who confirm such racialized suspicions is held as unimpeachable. Racialized marks of nationality—particularly attached to China, Russia, North Korea, Iraq, and Iran—become evidence of a potential threat, while the testimony of bodies is adjudged as reliable based on its tendency to confirm the established suspicion. Race, in this instance, is not dependent only on the reiteration of specific stereotypes attached to culture or phenotype. Racial profiling acquires momentum through the public mapping of uncertainty in a globally interconnected world by asserting the potential for nefarious motives among those world regions tainted by association with political Islam or communism. These logics intensify as rumor, speculation, authority, and experiential knowledge generated through laboratory research and government threat scenarios help transit the virtual body of variola across the public/secret divide.

The present section, which discusses the writings of Judith Miller and Richard Preston, examines how variola virtually transits through figures of racial suspicion, mobilizing unknowability into the affective intensity of biological threat. Judith Miller is the former *New York Times* reporter, more recently employed by a number of conservative news outlets and think tanks, whose yellow journalism on Iraq's supposed WMD was repudiated widely in the mainstream and Left media following the 2003 Iraq invasion and during the prosecution of I. Lewis Libby in the Valerie Plame affair. Before the scandal, Miller had years of contact with biodefense officials. She reports being "mentored" by William Patrick III, the top US bioweapons engineer of the 1960s.[42] In October 2001, Miller, William Broad, and Stephen Engelberg, all investigative reporters at the *Times*, published the book *Germs: Biological Weapons and America's Secret War*, which was the first inside account of the history and development of the US biodefense program. Drawing heavily on interviews with defense analysts and weapons researchers within the US military, *Germs* portrays a biosecurity apparatus at war with itself over the proper extent and reach of state research into potentially devastating biological agents. Contrasting Nixon's official dismantling of the US offensive bioweapons program in 1969 with the clandestine continuity of the Soviet bioweapons program throughout the Cold War era, the authors suggest that US biodefense policy oscillated between conciliation and dread over the potential of Soviet and post-Soviet bioweapons proliferation. *Germs* was written before the 9/11 attacks and achieved its top nonfiction spot on the *New York Times* bestseller list just after the anthrax scare in October 2001, when Miller herself received an envelope containing white powder in her mail at the *Times*.

For the authors of *Germs*, repeated examples of the government overestimation of biothreats did not detract from the general conclusion that "the threat of germ weapons" was "real and rising," spurred by the inevitable proliferation of biological "knowledge."[43] Geopolitical changes—ranging from the fall of the Soviet Union to the end of South African apartheid and the globalization of the information economy—meant that more knowledgeable experts were on the open market, willing to sell their services.[44] While *Germs* received mostly positive reviews, Simon Wessely, a British army advisor, pointed out one of the book's rhetorical excesses: it took past biological attacks that had failed to arouse public fear and retroactively invested them with the terrifying potential of mass death.[45] The book's representation of Saddam Hussein's Iraqi government does not focus on Saddam's personality or motives; Miller, Engelberg, and Broad do not attribute Saddam's pursuit of weapons to ideology or to a personal psychological profile. Yet they follow the neoconservative logics that separate Saddam and other apparent rogues from a universe of normative sovereign behavior, where nation-states act in a calculated manner depending on rational assessments of force and cost. They thus accept that Saddam should be singled out as a figure of excessive uncertainty, threatening to undo the rational calculus underpinning the nation-state system's common protections against illegal uses of force. In addition to disavowing the continuing usage of depleted uranium munitions and other unconventional weapons by the United States, this view capitalizes on the sense of risk already circulated by neocons and the security state around the figures of the rogue leader and the Islamic terrorist.[46] The authors directly refer to Miller's later source on Iraq, I. Lewis Libby, in order to suggest the growing threat of bioweapons despite any hard evidence: "Where was the next hotspot? What could be done to head off trouble? What were policy makers overlooking as they coped with the rush of day-to-day crises? . . . Libby told colleagues that intelligence analysts had an unfortunate habit: if they did not see a report on something, they assumed it did not exist. Or, as another veteran intelligence officer put it, absence of evidence is not evidence of absence."[47]

Germs' publication just after 9/11 made it an immediate success, and it was spun off into a PBS special on bioterrorism. It was followed by a series of high-profile news reports by Miller that staged the neocons' fears of Saddam in more direct, seemingly authoritative statements. Alexander Cockburn has, in retrospect, accused Miller of attempting to sell copies of her book through her yellow journalism on Iraqi chemical, biological, and nuclear weapons; however, Miller's position at the crossroads between state publicity and secrecy is more significant than Cockburn's generic accusation of a profit motive.[48] In the paranoid worlds of biodefense, rumor, speculation, and the staging of fear in the laboratory allowed threats to virtually transit across bodies and spaces, appearing legible in ways that convinced not only Miller, but also the *Times* editorial board. Bioweapons make potent affective weapons for the post-9/11 security state, as their attributed potential for destruction makes potential harm overwhelm calculations of probability.

The increasingly ominous headlines warning of Iraqi bioweapons signaled the risks of military inaction. Miller's key article for the *Times* concerning the potential for Iraq to possess and weaponize smallpox was published on December 3, 2002, at the same time that Donald Rumsfeld was publicly pushing the Smallpox Vaccination Program.[49] Miller's report expands on an earlier piece by Barton Gellman published in the *Washington Post* airing administration allegations that Iraq, North Korea, Russia, and France held covert stocks of variola.[50] (Gellman inexplicably escaped criticism over his yellow journalism in contrast to Miller's abrupt public fall from grace.) Both stories advertised administration deliberations over the smallpox program. Presenting justifications for this expensive program that was widely denounced by public health officials, Miller's article focuses on a government informant who alleged that Iraq obtained smallpox virus from a Russian scientist, Nelja N. Maltseva, whose documented travel to Iraq happened thirty years earlier. The scenario alleging the sale of smallpox to Iraq by Maltseva, a respected member of D. A. Henderson's smallpox eradication effort, emerges from a string of unconfirmed speculations that are pieced together in the article: a disputed theory that a small outbreak of smallpox in Kazakhstan in the 1970s was the result of open-air biological weapons testing; that Maltseva returned to Iraq in 1990; and that Maltseva would have had access to particular strains of Soviet-held smallpox at that time. Miller's reporting relies heavily on government sources retelling old intelligence derived from untrustworthy defectors, leaving out other statements by the same sources alleging that Saddam had destroyed remaining stocks of unconventional weapons. (These excluded claims were confirmed in a postinvasion review of the Iraq Survey Group.)[51] Her reports emphasize the inevitability of uncertainty itself, helping to build the neocons' case against inspections: "Verifying Iraq's assertions that it has abandoned weapons of mass destruction, or finding evidence that it has not done so, may not be feasible, according to officials and former weapons inspectors." As such, Miller relates, "The White House is expected to announce that despite the risk of vaccine-induced illness and death, it will authorize vaccinating those most at risk in the event of a smallpox outbreak—500,000 members of the military who could be assigned to the Middle East for a war with Iraq and 500,000 civilian medical workers."[52]

If Miller's reporting has been criticized due to her cozy relationships with the likes of Libby, the bioweapons journalism and fiction writing of Richard Preston are not so easily aligned with the Bush administration. Preston, a *New Yorker* contributor, is known for his best-selling nonfiction account of Ebola emergence, *The Hot Zone* (1994), which inspired the popular film *Outbreak*. After *The Hot Zone*, Preston turned his interest in biosecurity into a trilogy, completing *The Cobra Event* (1996), a novel exploring a fictive engineered virus attack on New York, and *The Demon in the Freezer* (2002), a suspenseful and research-heavy journalistic account of smallpox and anthrax weaponization threats. The original literary leaker crafting suspenseful scenarios of bioattack, Preston counts Steven Hatfill and Joshua Lederberg among his sources.

Preston marketed his "Dark Biology" trilogy as a journalist's experiment with the literary tool of suspense in order to advance public knowledge of the threats posed by advanced bioscience to international security. In *Germs*, Miller, Engelberg, and Broad claim that in 1997, a scientific advisor convinced Bill Clinton to read Preston's *The Cobra Event*, which caused him to look seriously at bioweapons preparedness. Having devoured Preston's book in one night, the story goes, a frightened Clinton embarked on a bungled attempt to establish bioterror response teams within the National Guard. Preston's book had been promoted in defense policy circles by Richard Danzig, undersecretary of the navy, who believed that "the military . . . was largely detached from biology" and that "without a visceral, direct sense of what the weapons could do to an enemy, military officers lacked compelling reasons to invest in vaccines, detectors, or better protective suits."[53] In addition to Preston's work, both I. Lewis Libby and Steven Hatfill themselves wrote outbreak novels, Libby's a deeply orientalist and pornographic portrayal of a smallpox outbreak in Japan and Hatfill's an unfinished narrative concerning an Iraqi bubonic plague strike at the White House.[54]

Suspense offers a privileged model for exploring risk in the fictions of literary leakers because it generically contains uncertainty in its plot resolution; the paradox of suspense is that it propels narrative through movement toward expected resolutions rather than actual uncertainty or unpredictability.[55] *The Cobra Event* does, at the outset, offer a swift and suspenseful narrative of biological terrorism, followed by a lengthy and overly detailed state response meant to outline potential missteps and systemic vulnerabilities as an engineered epidemic sweeps just past the grasp of public health and law enforcement officials. As cases of a mysterious disease transform a handful of New Yorkers into leaky, seizure-prone aggressors who cannibalize their own faces, two rogue US members of the Iraq UNSCOM delegation conduct a surprise inspection that nearly catches a mobile Iraqi bioweapons laboratory as it speeds away; the female Russian scientist left behind at the scene later appears at a suspect biomedical company in New Jersey, part of an international syndicate that seeks to develop bioweapons. While Russia and Iraq are linked to this shady organization, it is ultimately a misanthropic US scientist, aiming to control human population, who himself goes rogue from the organization and releases the experimental bioweapon without their knowledge. The genetically engineered pathogen, a combination of smallpox, rhinovirus (common cold), and nuclear polyhedrosis, proves difficult to identify. The scramble by CDC, FBI, and military officials to identify, diagnose, and contain the pathogen while catching the perpetrator fumbles on several counts: failure to effectively quarantine contacts; difficulty maintaining biosafety containment in an improvised forensic lab; dense population concentration in New York; ineffective communications; and a maze of subway tunnels that allow the perpetrator to nearly escape.

The lesson, spelled out in Preston's preface, is that "hope is an expensive commodity. It makes better sense to be prepared."[56] Preparedness is the antidote to catastrophic risk. Of course, since preparedness rests on the affective generation of uncertainty, it never actually mitigates suspicion; rather, it constantly regenerates it. Preston's book plays with the distinctions between fiction and nonfiction, public knowledge and state secrets; in the lengthy historical interchapters, Preston describes the histories of US bioweapons testing in the Pacific at Johnson Atoll and inspections of Russian and Iraqi programs, drawing heavily on interviews with unnamed government officials. Preston argues that while the story is fiction, "the historical background is real, the government structures are real, and the science is real."[57] Yet the role of Iraqis and Russians in the fictionalized international syndicate is opaque; the investigators never fully describe their role, leaving open uncertainty combined with assertions from American inspectors that they "just have a feeling" of Iraq's duplicity. The narrator adds without explanation, "There were indications that the Iraqi bioweapons programs were very much alive."[58] Endless uncertainty in the narrative manages to predictably narrow to conventional enemies in these racial geographies of suspense.

The particular indications of Iraqi duplicity in the book appear to be SBU details that line up closely with what was disclosed six years later in Colin Powell's speech to the United Nations. Powell presented animated diagrams of Saddam's alleged mobile laboratories, complete with imprecisely labeled fermenters and mixing tanks (bioreactors), which would open the possibility of manufacturing either viral or bacterial agents (figure 4.1). Powell listed a dozen potential agents, concluding the list with the most ominous claim that Saddam had "the wherewithal to develop smallpox." A disgraced Powell later

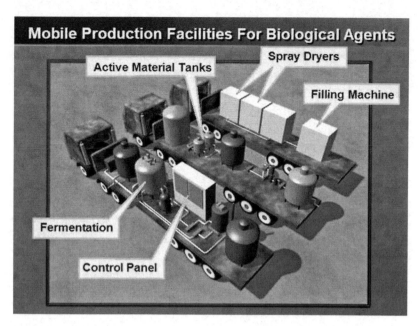

Figure 4.1 Mobile production facilities for biological agents, slide presented by the United States to the United Nations Security Council, February 5, 2003.

explained that the mobile labs accusation was included because it was "dramatic" and blamed the CIA for the erroneous claim.[59] Yet long before the CIA received its dubious claim in 1999 regarding mobile labs from an Iraqi defector to Germany, Rafid Ahmed Alwan al-Janabi, it had imagined that Iraq would follow the US example and establish such mobile weapons production. The United States had made labs on wheels for manufacturing anthrax during the Cold War program, as a backup in case Soviet attacks destroyed primary bioweapons capabilities. Then, in 1992, chief weapons inspector Scott Ritter proposed to the UN that Iraq could maintain such labs to evade the UN mission. In 1995, UNSCOM inspectors proposed the scenario to another Iraqi weapons program defector, who confirmed (self-aggrandizingly) that he had aired the idea but claimed the labs were never built.[60] In the echo chamber of US biodefense, the state's own preparedness for biowarfare made mobile labs intelligible; a series of leading questions to defectors seemingly confirmed the idea; and, ultimately, the circulation of racialized rumors "transform[ed] every truck in Iraq into a potential WMD laboratory."[61] Later, Ritter made himself into a hero of the antiwar movement despite planting the seed of the erroneous accusation used by Powell to justify the war.

The Return of Smallpox: Dark Winter and Uncertain Asias

When the *Germs* authors, following the navy's Richard Danzig, seek a "visceral, direct sense of what the weapons could do to an enemy," they call for the generation of forms of representation, like the suspense genre, that affectively model uncertainty in order to channel it into comprehensible form. Colin Powell attempted to produce a direct sense of threat by displaying in his hand a small vial of simulated anthrax powder to the assembled world leaders at the UN (figure 4.2). This performance highlighted the contrast between the small scale of the threatening object and an imagined scenario of mass death, one targeted at a major security institution and the many state officials gathered at the meeting. It also capitalized on the tautological suggestion that since Powell had himself managed to bring a suspicious material to the gathering, uncertainty and insecurity were systemic.[62]

By the time of the Iraq invasion, a number of bioterrorism exercises had been staged for municipal and federal officials in order to systematically provide other entries into this direct sense of vulnerability. It would be appropriate to think of these scenario-based exercises as a peculiar new subgenre of security theater in the twenty-first century. Andrew Lakoff explains that the scenario-based exercise was developed by military planners during the Cold War in order to apply "the method of imaginative enactment to the generic crisis situation to generate knowledge about internal system vulnerabilities."[63] Applied to public health circles for the first time in the 1960s, exercises relied on casting experts as actors who would receive information on an emerging threat and respond accordingly. The scripts of such exercises rely on staging not the probability of a threat, but the plausibility of it and the particular capacities for response. Like role-playing games, the scenarios are meant to be player driven, with multiple outcomes imagined in the scripts. (In the process, other alternatives, like inaction or multinational action, are closed down.) The emphasis on worst-case scenarios is evident in the staging of the largest of these exercises to date, TOPOFF, which was carried out in three urban centers in May 2000. The authors discuss the necessity of including nonenacted "notional" events in the scenario such as riots, mass casualties, and hospital runs, which, due to their mass scale and catastrophic form, cannot be realistically staged within budgetary, logistical, or spatial constraints.[64] Confronted with the text of life-altering emergencies, the exercise produces predetermined evidence of the vulnerabilities of vital systems to catastrophic events. State actors are placed at the scene of the action and faced with notional crisis

Figure 4.2 Colin Powell holding a vial of a mock bioagent at the United Nations Security Council, February 5, 2003. UN Photo by Mark Garten.

events that require their response. The progression of the notional catastrophe across the time of the crisis scenario generates the knowledge and feeling of vulnerability regarding a host of linked risks. The scenario is meant to produce a virtual reality of disaster.

In June 2001, the best-known bioterror exercise, Dark Winter, was staged at Andrews Air Force Base. Written by biosecurity experts Tara O'Toole and Thomas Inglesby, the scenario stages the National Security Council (NSC) response to a coordinated smallpox attack on Oklahoma City, Philadelphia, and Atlanta. Eighteen government officials and journalists played the scripted roles, while another fifty biosecurity officials watched. In the script, the authors turn to a familiar set of sources. They cite Judith Miller and William Broad to posit that countries other than the United States and Russia hold clandestine stocks of smallpox; turn to Ken Alibek's book to argue that Russia has ongoing capacity to produce hundreds more tons of variola annually; and reference William Patrick to claim that one gram of smallpox could infect one hundred people in an aerosol attack. Staging Iraq and North Korea as unpredictable enemies, and Russia as their silent partner, the scenario context focuses on East and West Asia as the geographic ground of uncertainty:

> The year is 2002. The United States economy is strong. Tensions between Taiwan and the People's Republic of China are high. A suspected lieutenant of Osama bin Laden has recently been arrested in Russia in a sting operation while attempting to purchase 50kg of plutonium and biological pathogens that had been weaponized by the former Soviet Union. The United Nations' sanctions against Iraq are no longer in effect, and Iraq is suspected of reconstituting its biological weapons program. In the past 48h, Iraqi forces have moved into offensive positions along the Kuwaiti border. In response, the United States is moving an additional aircraft carrier battle group to the Persian Gulf.[65]

As the NSC meets to discuss growing tensions in the Gulf states, the president announces a confirmed case of smallpox in Oklahoma City, and the presumption that this must be a bioterror attack. From here, a number of notional events unfold over the course of three NSC meetings and two weeks, demonstrating the state's supposed inability to act in the midst of uncertainty. As the scenario unfolds, politicians and analysts, including James Woolsey, Sam Nunn, and David Gergen act out the scripted posts on the NSC, while five reporters, including Judith Miller herself, portray journalists responding to the crisis. As Lakoff explains in his reading of the scenario, the initial questions are technical but carry grave political weight. The participants are asked to decide on whether to mobilize a limited supply of vaccine, and the epidemic quickly erupts, overwhelming hospitals, vaccine delivery, and law enforcement as panic sets in. By the third meeting of the NSC, 16,000 US Americans have smallpox, 1,000 have died, and estimates suggest that in the coming weeks, the epidemic could infect 3 million, killing one-third (echoing the inflated proportion of deaths estimated among American Indians in the Spanish conquest). Staging the lack of knowledge at the NSC in the face of an expanding epidemic, the conversation quickly turns to rush efforts to expand vaccines as well as enhanced spatial control of disease contacts; participants discuss militarized quarantine and the federal authority to deploy the National Guard to control transborder movement.[66] As in *Outbreak* and *The Hot Zone*, the exercise normalizes the strong hand of the state. On the Dark Winter website, the "Findings" section gives an approving nod to former Yugoslavian leader Joseph Tito for responding to a single case of smallpox in 1972 by instituting a national quarantine and forcibly vaccinating the entire population.[67]

This state militarization turns overseas as well, as the unseen terrorist sends a note calling for the pullout of the US military from the Persian Gulf; the final scene of the scenario cites an Iraqi defector as claiming the state's involvement in the attacks, even as Saddam denies wrongdoing and threatens retaliation if attacked. The scenario plays again with the problem of uncertainty. How can the state respond when the epidemic begins to fan out before the NSC even convenes? How can 12–15 million doses of smallpox vaccine protect a population of nearly 300 million? What decision calculus should be used to decide on emergency measures, both domestically and internationally? How should Iraqi duplicity be addressed? Given that none of these scenarios offer complete, certain information to the participants, the exercise manifests an urgency that suggests militarized preparedness and preemption; as the threat is imagined to escalate uncontrollably, killing thousands within days, no response seems too draconian. These logics are drawn out in responses to the exercise, which were aired in six hearings on Capitol Hill. For Sam Nunn, who played the US president, the scenario demonstrated, "You're in for a long term problem, and it's going to get worse and worse and worse and worse and worse." This means "cooperation from the American people. There is no force out there that can require 300,000,000 people to take steps they don't want to take." For James Woolsey, the top neocon in the Clinton administration, bringing questions of health inside "the framework of a malicious actor" suggests "we are in a world we haven't really been in before."[68] These statements indicate the intensification of fear produced by the form of an exercise in which notional crises inevitably get "worse and worse and worse and worse and worse."

This, however, is not how the authors describe the logic of the scenario. They suggest that the exercise itself reveals specific structural problems in the security apparatus that can be remedied. Even though the scenario is determined to be catastrophic in advance, they take its anticipatory logic as the simple expression of everyday risk that must be incorporated into security planning: "Leaders are unfamiliar with the character of bioterrorist attacks, available policy options, and their consequences. . . . The lack of sufficient vaccine or drugs to prevent the spread of disease severely limited management options. . . . The US health care system lacks the surge capacity to deal with mass casualties. . . . The individual actions of US citizens will be critical to ending the spread of contagious disease; leaders must gain the trust and sustained cooperation of the American people."[69] However, all of these warnings had been articulated previously by other biosecurity officials. The real knowledge produced through the scenario is affective—imaginative enactment, especially surrounding those violent notional events that escape full representation, animates variola as incitement to vigilant preparedness in the form of vaccine stockpiling, martial law, and the enactment of yet more scenario-based exercises. The entire scenario is based on a certain mapping of risk, one that centers on West Asia (the Gulf states, Iran, and Afghanistan) and East Asia—figured as regions that are economic or security threats contrasting with the fictionalized innocence of the United States. Despite Russia's origination of the smallpox in the scenario, other actors are more threatening. China appears belligerent toward the United States and appears to be mobilizing bioweapons against Taiwan. International terrorist organizations have shadowy ties to Russia and Iraq. The very fact that like all states, these targeted Asian states are not transparent—they insist on a monopoly on state secrets and state violence within their borders—means that uncertainty proliferates in international relations.

The proliferation of uncertainty staged in the scenario puts pressure on participants. This is especially true of Dark Winter's staging of news clips reporting on breaking events and circulation of pictures to identify smallpox victims. Each clip includes a graphic introduction that juxtaposes images of a pockmarked brown-skinned baby with smallpox, an illustration of a double helix, and a scene of police in riot gear pointing rifles, recalling the close association of Asiatic contagion with militarized quarantine in the case of Hansen's disease one century earlier (see chapter 1). One news brief breaks with the TV news format as it announces a "special report on the deadly effects of small-pox," only to wash out the audio and leave the viewer with silent scenes depicting bloody and ulcerated bodies, brown and thin, the camera panning over still legs, torsos, and faces. These visual cues stage the spectacle of debility and death caused by variola. The realistic images of suffering and death balance the speculative character of geopolitical, diplomatic, public health, and epidemiological factors in the scenario. Like the videos, the still shots distributed to the simulated NSC meetings also focus on brown, pockmarked bodies, in this case staging the progression of the disease among children. A final shot shows a notably lighter-skinned male body, also covered with smallpox rash, as the "first victim" of the disease in Oklahoma City.

All of these images of the debilitating effects of smallpox are taken from the archives of the smallpox eradication campaigns in the 1970s. Most of the over 1,000 photos in these archives were taken in India and Bangladesh, though a significant portion of images come from Africa. Rendered alien by both the disfiguring rashes and hemorrhaging of the disease and the grainy reproduction of old photographs, brown bodies provide the visual evidence of the mutant hybridity of human and variola. The particular images used in the script are adapted from the WHO smallpox recognition cards, which were displayed in public gatherings at hospitals, schools, and homes during the eradication (figures 4.3 and 4.4). These grainy amateur photographs of brown bodies from the local area were, in their intended use, displayed as a practical guide for the public to identify and report cases to health authorities in the defense of their own communities; in their appropriation for the Dark Winter exercise, they are removed from this context and communicate a racialized, border-crossing contagion that proliferates within the body as days progress. The white index case realizes the transmission of the alien disease from dark bodies to the domestic settler body. Race maps the return of a disease that unevenly impacted the globe back to America, its ground zero.

Dark Winter draws affective power from the uncertainty attributed to inscrutable Asian bodies, to primitive disease, to hypermodern research, and to emergent economic and military power that threatens the mythic Pax Americana. As in the case of Hansen's disease during Hawai'i's annexation debates, Asia signals the threat of rising globalized networks to a precarious US empire. Dark Winter works to reanimate variola, to envision its power to sweep across the contained borders of Cold War empire. As it intensifies suspicion around figures of racial threat, it also attempts to justify state action by activating the sense of vulnerability.

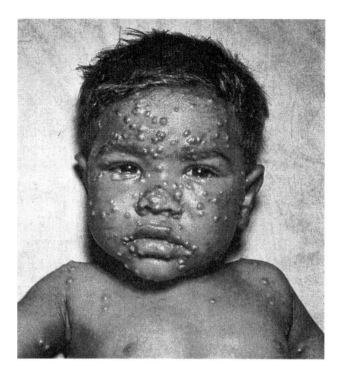

Figure 4.3 Smallpox identification card, courtesy World Health Organization. Originally published in F. Fenner et al., *Smallpox and Its Eradication* (Geneva: World Health Organization, 1988), 499.

Figure 4.4 Indian schoolchildren viewing the smallpox recognition card, courtesy World Health Organization. Originally published in F. Fenner et al., *Smallpox and Its Eradication* (Geneva: World Health Organization, 1988), 501.

On the Limits of Intervention

Following the lead of Preston's *The Cobra Event*, the first antiwar novel set in post-Saddam Iraq written by an Arab author conjures the threat of a mysterious virus modeled on the destructive capacity of recombinant smallpox. Yasmina Khadra's *The Sirens of Baghdad* centers on the alienation produced among Iraqis by the US occupation from the vantage of a desert village in Iraq. It traces how the social transformations and pervasive acts of violence unleashed by American forces work to radicalize the narrator, a young Bedouin man whose family home is indiscriminately raided during a security sweep. After working his way up the ranks of the criminal organization of a Baghdad strongman, the narrator graduates into participating in a larger militant organization's attempt to exact revenge on the imperialist West. Not knowing his mission, he is only told that he will be pivotal in *"the greatest operation ever carried out on enemy territory, a thousand times more awesome than the attacks of September 11."* Suggesting that this desire for "the whole planet . . . to go up in smoke" is political rather than cultural or religious, the novel proceeds to document the most destructive attack imaginable. It is the spread of an engineered virus that the narrator will seed indiscriminately in London; the virus is fantasized to speed across transit networks to other sites in the West, as described by the microbiologist of the terrorist organization:

> Your mission will then consist in riding the subway and going to train stations, stadiums, and supermarkets, with the goal of contaminating the maximum number of people. Particularly in train stations, so the epidemic will spread to the other regions of the kingdom. The phenomenon propagates with lightning speed. . . . This new [virus] is unique, and we alone have the knowledge that will be required to stop its further spread. And our intervention will require compliance with certain conditions. This is an unstoppable mutating virus. A great revolution. . . . It is *our* ultimate weapon.[70]

Khadra, it turns out, is the pen name of Mohammed Moulessehoul, a former Algerian army officer who wrote under a woman's name in order to escape censorship by the Algerian military. His identity was revealed after his migration to France, which caused a small furor in French literary circles since his work had previously been hailed as an authentic Arab woman's critique of both the US war on terror and the patriarchal violence of Islamic militancy. Khadra's novel situates the fantasies behind militant violence within a singular space of violence united with the US imperial occupation rather than as a clash of civilizations. The anti-imperialist fantasy of bioterrorism is actually appropriated from empire itself; the blowback here is determined by the very form of the suspicions leveled at Iraq by the neocons. While Khadra's perspective as an officer charged with quashing Islamist rebels in the Algerian civil war is actually closely aligned with US security agendas, his vision of blowback queries whether the West's obsessive focus on bioterrorism might actually work to constitute its threatening object rather than control it. This is a scenario of regular speculation within the biosecurity apparatus, as officials agonize over how to control published biosecurity research.

From a different vantage point, British playwright David Hare's drama *Stuff Happens* attempts to explain the Bush administration's decision-making process on the Iraq War by revealing what Hare sees as the rifts and struggles among different ideological viewpoints in the administration. Colin Powell and Condoleezza Rice lose the internal administration debate depicted by Hare, steamrolled by the neocons who have Bush's ear. Unlike Khadra, who adopts the logic of blowback to turn speculation of smallpox risks on their head, Hare presents the war as an effect of human dramas and ideological battles within the enclosed worlds of the security state and multinational diplomacy.[71] While this is effective in questioning the presumption of a single set of economic interests guiding the logic of war, its liberal vision of Powell as a tragic victim of the neocons obscures the strong affective ties linking race, speculation, and species in the exercise of biopreparedness. Hare thus risks turning attention away from how Powell's performance at the United Nations worked to affectively signal, not just to explain, terrorism's catastrophic potential. Against the uncertain global maps of risks, the small weapon adds the relief of fine detail: the mobile laboratory hiding in plain sight; the terrorist meeting in the open; concealed knowledge circulating between scientists and rogues; the casual dispersal from a vial, a cough, or a backpack of engineered infections along global routes. In the queer, ever-shrinking scales of bodily contact from which hidden objects erupt into mass violence, emerging projects of imperial security find the next big threat to US empire, which in the fantasies of American exceptionalism is always taken as a threat to the entire world, humanity itself.

In its defense of the Bush doctrine, the Project for a New American Century emphasized that, even if no weapons were ultimately uncovered, and even if there was no imminent threat to the United States in Saddam's Iraq, the war was carried out according to a coherent logic of preemption. The project quotes George W. Bush: "We must take the battle to the enemy, disrupt his plans, and confront the worst threats *before they emerge*."[72] Anticipating catastrophic biological and chemical attacks, the project stresses the bipartisan imperative to act in the wake of 9/11:

> On this question, the record is clear. Both administrations assessed the threat in similar terms. For example, in September 2002, Secretary of Defense Donald Rumsfeld asked CBS News viewers to "imagine a September eleventh with weapons of mass destruction. It's not three thousand—it's tens of thousands of innocent men, women and children." Similarly, Clinton administration Defense Secretary William Cohen stated in a televised Pentagon press briefing that the "UN believes that Saddam may have produced as much as 200 tons of VX, and this would, of course, be theoretically enough to kill every man, woman and child on the face of the earth. . . . We face a clear and present danger today. . . . [The] terrorists who bombed the World Trade Center in New York had in mind the destruction and deaths of 250,000 people that they were determined to kill."[73]

The fact that the neocons were widely discredited over their weapons claims does not mean that they were unsuccessful in their efforts to bolster an anticipatory and militarized state approach to biosecurity that continues to influence state practice. Whether articulated through biological, drone, informational, or environmental warfare, these logics have not been disrupted by the global condemnation of the Iraq War. With 9/11 looming large over every US president, preparedness is only likely to take on a greater share of budgetary commitments and research within the security apparatus, even as austerity forces place pressure on social welfare programs like national health care that are far more likely to bolster public health. For the large and growing antiwar and antiracist movements contesting the war on terror to effectively formulate new agendas on the security state, it will be necessary to take on this significant structural and affective dimension of the government of species.

It is conceivable, though highly unlikely, that states will face weaponized smallpox attacks in the coming decades. In the event of a successful attack, it is unclear that smallpox would spread uncontrollably as imagined in the apocalyptic virgin immunity scenarios of US defense planners. Regardless of the probability of attack, the security state has now stockpiled vaccine for 300 million US residents at a cost of $500 million, despite the fact that Henderson's eradication campaign was successful in the 1970s even though it vaccinated only a small percentage of at-risk populations. Preparedness reconstitutes itself against the plausibility of smallpox weaponization in a chain of interventions into public health, military, and media domains. In the process, it upends normative risk-benefit thinking about public health priorities, catalyzing further health interventions, imagined threat scenarios, and narrative and informational warfare carried out by the security state. The very suspended animation of variola, its uncertain future, added affective potency to its virtual transit, intensifying threat through both laboratory animals and racial figures like the vanishing Indian and the Islamic rogue/terrorist. In the process, the settler-colonial homeland was affectively catalyzed as increasingly vulnerable to uncertainties mapped upon Asia, interlinked with the mutant potential of a zombie virus that jumps across the divides between laboratory and environment, state secrecy and public culture. The result was a lengthy occupation of Iraq, a new high-tech vaccine delivery system for the domestic population, and a heightened interest among public health officials in militarized disease interventions in the new century.

CHAPTER 5

Refugee Medicine, HIV, and a "Humanitarian Camp" at Guantánamo

I'd been in prison, a tiny little cell, but crammed in with many others, men, women, and children. There was no privacy. Snakes would come in; we were lying on the ground; and lizards were climbing over us. One of us was bitten by a scorpion. . . . There were spiders. Bees were stinging the children, and there were flies everywhere: whenever you tried to eat something, flies would fly into your mouth. . . . I said to myself, come what may, I might well die, but we can't continue in this fashion. We called together the committee and decided to have a hunger strike. Children, pregnant women, everyone was lying outside, rain or shine, day and night. After fifteen days without food, people began to faint. The colonel called us together and warned us, and me particularly, to call off the strike. We said no. At four in the morning, as we were lying on the ground, the colonel came with many soldiers. They began to beat us—I still bear a scar from this—and strike us with nightsticks. . . . True, we threw rocks back at them, but they outnumbered us and were armed. They then used big tractors to back us against the shelter, and they barred our escape with barbed wire.
—YOLANDE JEAN, quoted in Paul Farmer, *Pathologies of Power*, 1996

Guantánamo, everyone agrees, is an animal—there is no other like it.
—SUPREME COURT JUSTICE RUTH BADER GINSBURG, oral arguments in *Rasul v. Bush and Al Odah v. United States*, 2004

In 1991, Yolande Jean was a Haitian democracy activist who feared arbitrary violence and imprisonment in the wake of the military coup that deposed President Jean-Bertrand Aristide. Yet Jean's story of imprisonment and abuse following the coup comes not from Haiti, but from her experience in the hands of the US Coast Guard and Immigration and Naturalization Service (INS). Jean was one of approximately 35,000 Haitian refugees apprehended in the

Caribbean Sea by the US Coast Guard following the military coup, which was tacitly supported by the elder Bush administration. The 10,000 refugees who were not immediately returned to Haiti by the Coast Guard were transported to the US naval colony outside Guantánamo, Cuba, where they were screened for human immunodeficiency virus (HIV). The 269 Haitians who tested positive, including Jean, were segregated and imprisoned in the world's first HIV concentration camp during the years 1991–94.

Jean's story, told to the health activist and anthropologist Paul Farmer, is striking because of its account of military violence and unsanitary conditions at a camp for medical prisoners. It also depicts unwanted spatial intimacies that both physically threaten the body and symbolically mark the space of the camp as inhuman. As the refugees already live crammed together in a world with "no privacy," the snakes, lizards, scorpions, spiders, bees, and flies form a menagerie of threatening bodies that crawl over, sting, and enter the mouths of potentially immunocompromised refugees who lie exposed to the elements. Reversing Alfred Mouritz's localization of alimentary disgust in the mouth of the colonial Hansen's disease patient described in chapter 1, Jean figures the prisoners' mouths as the pathways of a state violence that accentuates brute force with the unwanted intimacy of bestial touch. With the image of live flies buzzing into the mouths of feeding prisoners, rendering the flesh of food and pest indistinguishable, Jean concludes that camp life is unlivable, inhuman. Even as Jean means to publicize the literal failure to contain refugee bodies from animals figured as pests, she also associates the state with an exceptional violence outside the law, an association of the sovereign and the beast bluntly captured by James Dieudonne, who was twelve years old when he spoke to the press about his experience of quarantine at Guantánamo: "It's not a place to play. It's for military and animals, not for a child to grow up."[1]

The US government claimed in court that Guantánamo was a "humanitarian camp" aimed at protecting the lives of HIV-positive refugees.[2] The *New York Times* ran a headline that depicted Guantánamo as an "oasis" for Haitians, reflecting the viewpoint of the military guide who escorted the journalist along the Pentagon-organized tour.[3] It is clear that such descriptions of the camp downplayed the violence of emergency quarantine. Nonetheless, to reject the cynical characterization of Guantánamo as an oasis does not mean that the state was incorrect to associate it with the practice of humanitarianism. In fact, I argue that in the post-*Jacobson* order of global health, carceral control and the management of the ill through logics of emergency intervention are entirely compatible with the liberal logics of humanitarianism. This is apparent given that even as many accounts of the Haitian HIV program by US lawyers, activists, and journalists presented sympathy for the tragic plight of the Haitians, their published accounts of the crisis also regularly excused the state's heavy-handed treatment of refugees given the administrative challenges they appeared to pose to the security state.

Yet more recent writings on the carceral quarantine of HIV-positive Haitian refugees at Camp Bulkeley, Guantánamo by Paul Farmer and Naomi Paik articulate a critique of the program assembled from the experiences and testimony of incarcerated Haitians.[4] Building on Farmer's and Paik's studies, this chapter explores how the association of HIV with Haitian refugee bodies helped to realize the fantasized emergency scenario of military quarantine that is repeatedly envisioned as the last line of defense against globalization's processes of disease emergence. Taking a form that was even more restrictive than that witnessed in Hawai'i during the annexation-era panics over Hansen's disease, carceral quarantine at Guantánamo reveals connections between fears of sexual and personal contact arising during the AIDS scare with long-standing US depictions of black (and specifically Haitian) bodies as dependent and impoverished. Exploring prisoner responses to the racialized management of life and death, this chapter attempts to reconstruct the biopolitical contestations of camp life at Guantánamo during the AIDS scare. I pay special attention to the interspecies entanglement of bodies—human, animal, and viral—within this carceral experiment, exploring the terms upon which Haitians responded to the media and the security state's racialized logics of risk constructing AIDS as an emerging, foreign disease. I argue that the biopolitical constitution of Haitian refugees as human-viral hybrids underwrote two types of state force: while, on the one hand, Haitians were written into national HIV policy as uniquely contagious transborder vectors in need of militarized exclusion, on the other, their bodies were made into zones for the exercise of a certain type of care that could be publicized as humanitarian to broader audiences in the United States and Haiti. This in turn generated an affective politics situated in the form of life itself, wherein the tortures of concentration, inadequate provision, and the failure to shelter ill Haitian bodies from the elements and from contact with the menagerie of species in the camp created the conditions for an organized Haitian hunger strike. Hunger striking challenged the logic of carceral quarantine by using deprivation and the specter of the imminent death of the captive to put pressure on the state's adjudication of Haitian claims and to publicize the lie that ostensibly humanitarian medical care constituted a benevolent intervention. This involved the first ruling to establish a legal standard for incarceration based on medical tests attempting to determine the capacity of the immune system itself. The logic of indefinite HIV detention signals the multiple positions of Haitians vis-à-vis the imperial security state—the rendering of the Haitian body as racial dependent; an excess of movement, contagion, and plasticity; a contested figure of the human; and a site for the exercise of state biological surveillance and care. The HIV-positive Haitians came into conditions of extreme biological and political vulnerability, and were simultaneously rendered as sentimental refugees and contagious human-viral hybrids in the world's first AIDS concentration camp.

At the end of the chapter, I draw broader lessons about the work of the racial security state and the return of indefinite detention to Guantánamo following 9/11. While HIV detention at Guantánamo responded to the specific contexts of US-Haitian transnationalism, it also shared formal parameters with the situation of present-day prisoners caught in the cynical abuses of the so-called war on terror. The return of hunger striking at Guantánamo in the last decade, as well as the military's attempts to advertise humane care there, display the ongoing tensions and linkages between health and security as governing universals of US empire. As the legal scholar and Guantánamo lawyer Muneer Ahmad writes, "Guantánamo is not simply a question of abstract legal memoranda or an exercise in line drawing, but a concrete project of state violence enacted upon, and realized in, the human body."[5] Further, I argue, this violence realized in the human body incorporates many populations, species, spaces, and institutions that converge to articulate both the space of incarceration and the regeneration of collective intimacies that contest indefinite detention.

US Empire, Haitian Refugees, and the World's First HIV Camp

Before the development of effective multidrug treatments in the mid-1990s, HIV diagnosis was commonly viewed as a death sentence.[6] Those diagnosed with it faced persistent stigma, especially if they were identified with one of the "4-h" risk groups: homosexuals, heroin users, hemophiliacs, and Haitians. Control efforts involved public attempts to transform sexual and health practices as well as the expansion of surveillance of bodies deemed at risk. Despite loud and repeated opposition from public health and immigrant rights organizations, the US government carried out a militarized program of containment aimed at Haitian immigrants. In 1991, by which time 2 million US Americans were already infected with HIV, Congress set its sights on maintaining the exclusion of HIV-positive immigrants and paid special attention to a class of immigrants from Haiti. This was an unusual step for the CDC to use national identity as a risk factor for a disease. In the midst of a large refugee flow coming north across the sea from Haiti, the government returned many directly, deemed ineligible as economic, rather than political, refugees. The 10,000 remaining who were "screened in" (meaning they were determined to have potentially valid asylum claims) were brought to Guantánamo for immigration and health screenings. The 269 Haitians who tested positive for HIV were segregated and held at a special detention camp called Camp Bulkeley. These prisoners were indefinitely detained and complained of inedible food, heavy-handed military policing, the forced labor of sanitary duty, shoddy shelter, and enforced medical care without proper information and consent. They meanwhile lacked freedom of movement and association, contained within the boredom, uncertainty, and unfamiliar environments of quarantine.

Because these refugees were undergoing processing for asylum in the United States, they could not be forcibly repatriated to Haiti; yet because they showed positive serostatus (the measured presence of HIV antigens in the blood), they ran afoul of the HIV exclusion policy that was unanimously passed by the US Senate in 1987. While the US attorney general had the legal authority to individually parole asylees and waive their exclusion, such provision was forcefully denounced by lawmakers who conjoined fear of Haitian economic dependency and contagion.[7] In three years of litigation, INS lawyers consistently presented both the refoulement of migrants (their preemptive return to Haiti before reaching US waters) and indefinite detention of HIV-positive refugees as twin efforts at national security and refugee relief: the United States wanted to discourage further migrants who risked their lives to cross the Caribbean Sea, while also providing humanitarian medical care for HIV-positive asylum seekers at a segregated site on Guantánamo. The US military hastily created an improvised quarantine in an area called Camp Bulkeley, where HIV-positive detainees were held behind barbed wire.

Even though the word *Guantánamo* today is synonymous outside of Cuba with the post-911 US-run military gulag, the base involves other spaces involved with a long-term military occupation dating to 1903. The relationships between the prison, the military occupation, and longer routes of colonial power and inter-Caribbean transit continue to shape the space of incarceration. Amy Kaplan's statement that Guantánamo is "the heart of American empire" needs to be elaborated in relation to a broader set of Atlantic and world histories that make it such a flexible site for the routing of US-Caribbean economic, military, and biological flows. Given the history and spatial organization of Guantánamo, it is not simply the ambivalent sovereign structure of US state apparatuses at Guantánamo—the fact of its legal nonjurisdiction despite a consistent occupation—that makes it repeatedly significant to US state power. Guantánamo involves complex imperial relations of economic, legal, medical, military, ecological, and political force. If Kaplan may have overstated Guantánamo's exceptionalism in the lengthy, planetary routes of US economic, cultural, and political dominance, she and other critics of US empire are right to point to Guantánamo's repeated utility as an occupied offshore site for population concentration and control.[8]

Descriptions of the architecture of Guantánamo by the lawyer Muneer Ahmad and former military interrogator Erik Saar help contextualize the forms of spatial segmentation that shape time and space within the small US colony at Guantánamo.[9] The US-controlled base sits upon a beautiful bay separated from the rest of Cuba by a live minefield. Provisions come in by plane and daily FedEx flights, and water comes from a desalination plant since the Cuban government cut off access to the domestic water supply. Free-moving soldiers and laborers use a ferry to cross, at scheduled daily intervals, from the residential side of the bay to the military camp and back. The residential side is also home to a long-term US military community living in tracked housing reminiscent of US suburbs. The town features a nostalgic commercial strip complete with all-American businesses like McDonald's, an A&W diner, and a Starbucks. The Filipino, Haitian, and Jamaican migrant workers who allow Guantánamo to operate within the low cost structures of the Caribbean labor economy share ferries with military police and US civilian personnel including families and US employees who operate institutions like a public school. (This mix of people allowed eleven Haitians to briefly escape from the HIV camp in 1993, passing as nonrefugee migrant laborers.)[10] The base itself is heavily guarded, and includes large open spaces as well as institutions including a base hospital and an airfield. Inside and outside the military base housing the infamous razor-wire detention camps, cars move at a sluggish pace to protect endangered iguanas.[11] The spatially divided, slow and regulated pace of human life at Guantánamo simultaneously suggests imposed military discipline (which compromises the work of media and lawyers who have limited time to visit), an advertised US state commitment to the protection of life and ecology, and a nostalgic production of Americana through the slow pace and segregated layout of the colony.

In addition to this direct territorial occupation, US influence in Cuba, Haiti, and the Caribbean more broadly has shaped the articulation of the AIDS crisis within the Haitian diaspora. C. L. R. James said in 1962 that "the Caribbean is now an American Sea," reflecting the strong hand of the navy in controlling migration and the influence of US capital in shaping tourism and other industries in the region.[12] This relates directly to deep inequalities in an era of globalization that make Caribbean states and economies dependent on outside tourists and trade. Thus HIV enters Caribbean space on the routes of these forms of transnational power. Paul Farmer argues that US tourism brought the virus to Haiti. Unfortunately, during the AIDS scare Haitians became objects of an erroneous racial epidemiology that posited that HIV made its zoonotic leap from monkeys to humans at the edge of the African rain forest and then spread from central Africa to Haiti to urban America.[13] The high prevalence of HIV among diasporic Haitians in New York in the mid-1980s brought about a CDC designation of being Haitian as one of the four risk factors for the virus. Paul Farmer critiques the racial misreading of epidemiological data behind this national and racial designation, which actually demonstrates that US tourists most likely brought the virus to Haiti in the late 1970s. The effects of the stereotyping were devastating: many Haitian immigrants lost their tenuous grips on employment in the United States and were subjected to profiling in schools.[14]

In 1991, antiretroviral treatment with azidothymidine (AZT) was still new, and HIV was still publicly viewed as a death sentence. Politicians continued to stoke fears of the 4-h risk groups (heroin users, Haitians, hemophiliacs, and homosexuals), as well as sex workers. In the mid-1980s, several US states including California and Texas took steps to quarantine persons testing positive for HIV, and proposals were floated for isolating sex workers to prevent the spread of disease. While isolation measures were ultimately rejected as national policy, the US Senate in 1987 unanimously adopted North Carolina senator Jesse Helms's proposal to add AIDS to the short list of diseases (primarily venereal diseases) on the CDC's immigrant exclusion list. Against the recommendation of the US Public Health Service, the Helms amendment reflected the moral taint publicly attached to the disease, in contradiction of expert opinion that there was only one disease contagious enough to warrant the exclusion of travelers and immigrants: tuberculosis. The fact that the CDC exclusion list (finally revised to remove HIV in 2010) consists largely of venereal diseases and Hansen's disease confirms its use in the service of moral stigma; Helms publicly conceived of HIV as divine judgment upon homosexuals, always already originating in "sodomy."[15]

If AIDS was commonly viewed as a moral taint within the United States, it also confirmed long-standing racial stereotypes about Haitians. The uniformly militarized response to Haitian refugees from the Carter through the Obama administrations reflects a central ambivalence in the US racial figuration of Haiti. On the one hand, Haiti represents a model of dependency. It appears plagued by instability that reveals not only the sorrows of poverty and inequality but also a kind of spiritual taint attached to repeated disasters and deferred hopes for liberation. On the other hand, Haiti is contained within a predictable colonial racial ordering that figures the uncontainability of black bodies as source of the persistent ungovernability of the island. The United States refused to diplomatically recognize Haiti from 1804 to 1862 because of the fear that black revolt was a kind of contagion that could spread to the US South. Vodou and zombies have been obsessively invoked by US Americans since the early twentieth century as representations of the failure of Haitians to observe proper cosmic order. This racial taint of Haitian disorder has been reinforced by repeated histories of occupation and servitude: the official de-recognition of Haiti, 1804–62; repeated US and European interventions (including the US military occupation of 1915–34); economic blackmail through colonial debts; enforcement of direct and indirect labor suppression; and the quiet Republican undermining of Jean-Bertrand Aristide, leading to two US-supported coups. Haiti is thus figured as a space where the impoverished are not only victims but also inevitably the agents of their own disorder and unfreedom. Haitian self-government, at the individual and collective levels, remains "unthinkable" in the words of Michel-Rolph Trouillot.[16] US posturing on Haiti involves advertising a myth of US humanitarianism as it disavows the roles of slavery and empire in shaping Haitian dispossession. Even in the aftermath of the 2010 earthquake, the Obama administration publicly advertised that the United States would allow the immigration of Haitian orphans, while it quietly readied tent camps at Guantánamo in the event of a repeat exodus of Haitians on the 535-mile journey by boat to Florida.

Nowhere is the figuration of dependency and ungovernability more apparent than in the image of Haitian boat people. Since the first arrival of the so-called boat people from Haiti on US continental shores in 1963, there has generally been little hope for official asylum, even though other maritime migrants from Southeast Asia and Cuba were broadly allowed entrance into the

United States on humanitarian grounds, and commonly granted humanitarian parole for excludable diseases.[17] In response to the early waves of Haitian boat people with the rise of Jean-Claude "Baby Doc" Duvalier in 1971, Jimmy Carter initiated the 1978 Haitian Program, which distinguished Haitians from all other potential asylum seekers by nationality; while Carter welcomed refugees fleeing communist states, and advertised the United States as the humanitarian savior of the Vietnamese, he attempted to block migration flows from allies and capitalist states that would threaten Cold War narratives of American exceptionalism.[18] Refugees faced militarized practices of refoulement that singled out Haitians but also drew Haiti into a broader program of border militarization against Mexicans and other potential refugees from allied capitalist states across the Americas. These migration outflows exhibited the lie that unfettered capitalism brought peace, freedom, and prosperity.

The United States disavowed its role in the production of Haitian authoritarianism, particularly under the father-son dictatorships of François and Jean-Claude Duvalier. In the 1960s and 1970s, Haitian migrants attempted to flee the twin scourges of dictatorship and neoliberal economic reforms, both underwritten by US policies trading support for the Duvaliers in return for US sweatshops and other economic expansion into Haiti.[19] Naomi Paik outlines the imperial contradictions underlying Haitian refoulement:

> Indeed, it is because the United States has fostered Haiti's unlivable conditions that it refuses to recognize Haitians' claims. Acknowledging the status of thousands of Haitians as refugees would be tantamount to admitting that the United States supported cruel dictators who murdered and terrorized their own people, an admission that would tarnish the US image as the world's primary defender of freedom and democracy. In broader terms such an admission would suggest the fallacies of Cold War ideologies contending that capitalism consistently aligns with democratic governance and is morally superior to any form of communism. Because the United States supported Haiti's (violently repressive) governments that serviced (exploitative) economic structures and Cold War ideologies, it made the Haitian refugee an impossibility—no person from the ostensibly democratic, capitalist state of Haiti could have a legitimate claim to asylum.[20]

The US Coast Guard interdicted and detained asylum seekers before they reached US waters. They were refused a number of national and international legal rights, including the right of nonrefoulement or nonreturn established by the United Nations in 1951; the right to habeas corpus or legal standing;

the right to an attorney during immigration proceedings; and the right of informed consent to medical procedures. While New York District Court judge Sterling Johnson's court order led to the closure of the camp in June 1993, the Supreme Court later upheld the Clinton administration's appeal. This effectively legalized the continued interdiction of Haitian refugees at sea, the state imperative of relabeling asylum seekers as illegal immigrants, and the right of the military to detain prisoners on Guantánamo.

The Camp Bestiary and the Force of Humanitarianism

Flying in the face of the public proclamation that Guantánamo was a humanitarian camp, the practice of concentration and the refusal of better medical facilities in nearby Miami increased threats to life and limb for patient-prisoners. The open-air prison camps repeatedly constructed as emergency detention facilities at Guantánamo accelerated the biological precarity of those quarantined through practices of concentration, deprivation, and exposure, making normative political contestation over camp life most often centered on whether the camp accomplishes the simple biological provision of life itself. As I explained in chapter 1, this very situation of deprivation quickly makes the camp into a site of contestation and reform, an emergent space of humanitarian intervention and the introduction of new disciplines of care. As such, the camp has an intimate if conflicted relationship to the expansion of liberal humanism under empire. The establishment of an ostensibly humanitarian refugee camp for HIV-positive Haitians at Guantánamo echoes this imperial linkage of displacement, concentration, and care.

Emerging diseases like HIV theoretically required major governmental responses that marshaled medical, public health, sanitary, and law enforcement powers. So where were the doctors at Guantánamo? There were only two military physicians stationed to handle the 269 stranded Haitians. For all the talk at the time of new frameworks and preventative measures for emerging viruses among public health officials, the state response to HIV retrenched some quite predictable segregated geographies of public goods. Immigrants held a symbolic role in mediating the geographic vision of disease emergence, as they were understood to move contagions across borders and were associated with microbial ecologies wherein disease could make zoonotic leaps from animals to humans. Smallpox eradicator D. A. Henderson, in a foundational essay on emerging disease response following AIDS, argued that a global disease surveillance system needed to target areas "where migrants and travelers from rural areas are found." Henderson wanted to focus on those liminal sites where humans brushed up against both a diversity of nonhuman species and the edges of global transport networks: the racially potent ecologies of fantasized emergence. Health centers needed to be located "in developing countries" and preferably in "more densely populated areas and those near the tropical rainforest."[21]

If military interdiction attempted to disrupt flows of emergence that it saw as swarming across US borders, Guantánamo placed potentially immunocompromised individuals in a space where they would be exposed to additional diseases; it increased disease risk at the borders of empire. It was one instance of a larger policy in which returning immigrants to home countries likely increased the spread of HIV because it took immigrants who had been exposed to US-occupied spaces with relatively high rates of HIV infections and deported them to home countries experiencing lower rates.[22] Yet in this particular instance, the INS simply wanted to indefinitely detain the refugees until they died. While the military provided two doctors to serve the HIV patients, it also deployed a force of ninety military police to contain Haitians living with the world's most feared virus. When the doctors requested airlifting out prisoners who were experiencing advanced symptoms of AIDS, they were rebuffed by the INS. An INS official bluntly stated, "They're going to die anyway, aren't they?"[23] Refugees were required to wear ID bracelets, were refused access to lawyers during their immigration screenings, and were confined in close quarters behind razor wire. In addition, refugees suffered inadequate food and shelter and an improvised medical order that treated prisoners carrying a life-threatening virus without established norms of informed consent; doctors called in by lawyers from Yale Law School uniformly complained that the carceral structure of the camp destroyed basic trust in the medical apparatus among the prisoners. The use of the birth control medication Depo-Provera without the knowledge or consent of women in the camp constituted a neo-eugenic assault.[24] An HIV diagnosis meant confinement, concentration, deprivation, and the targeting of all kinds of reproduction: sexual reproduction, to be sure, but also the reproduction of bodies, community institutions, and legal subjects.

Echoing Jean's description of the menagerie of animals coming into close proximity with those confined in the camp, prisoners' interviews and legal depositions repeatedly use metaphors of animalization to describe their dispossession and imprisonment. The camp structure had everything to do with redrawing the borders of species: there was the compromised status of human rights; microbial species were the object of state surveillance in medical screening; and Haitian refugees were subjected to difficult camp life in constant contact with various animals crossing the barbed wire. The carceral architecture at Guantánamo was highly improvised in form, with the hasty erection of nine different camps for housing prisoners. After being subjected to blood screenings, HIV-positive refugees were given conflicting information regarding their serostatuses, treatments, and futures in detention. When they eventually contested their incarceration, they were arrested, confined in small barbed-wire cells, and beaten by military police in full riot gear. Yolande Jean describes the violence following dizzying relocations from boat to screening center to camp:

> We had been asking them to remove the barbed wire; the children were playing near it, they were falling and injuring themselves. The food they were serving us, including canned chicken, had maggots in it. And yet they insisted that we eat it. "Because you've got no choice." And it was for these reasons that we started holding demonstrations. In response, they began to beat us. On July 18th, they surrounded us, arrested some of us, and put us in prison, in Camp Number 7. . . . Camp 7 was a little space on the hill. They put up a tent, but when it rained, you got wet. The sun came up, we were baking in it. We slept on the rocks, there were no beds. And each little space was separated by barbed wire. We couldn't even turn around without being injured by the barbed wire.[25]

The barbed wire, inadequate food and shelter, lack of telephone communications, intensive confinement, and excessive use of police force are common descriptions. As Paik observes, while the refugees' depositions for the district court case addressed the inadequacies of the living conditions—an issue that framed the court case and the refugees' public advocacy—they also refused the idea that "the refugees' humanity was reducible to their physical selves."[26] The indefinite time of the detention, the improvised conditions of spatial confinement, and the repressive police measures were narrated as a spiritual enclosure. Refugees experienced a suppression of shared community institutions, a crossing of the body with alien species, and feelings of boredom in which life itself seemed to stand still.

Haitian refugees' rendering of the camp as a site of aberrant interspecies intimacies—both literal and figurative—is striking. Haitians' consistent invocation of animal imagery in their accounts of mistreatment at Guantánamo depict gothic scenes of pestilent contact with reptiles and insects, as well as more generic comparisons to the lives of dogs, birds, and pigs. These images serve to both literally describe an unconventional ecology of incarceration and figuratively depict the spiritual assault on the incarcerated body. Jean recounts the inedible chicken with maggots, while the elected president of the prisoner association, Vilsaint Michel, describes food infested with "flies and worms." The improvised shelters—tents and shacks—failed to keep out the rain or shield detainees from the heat; they were, for Michel, like "a bird's house where you would put pigeons."[27] According to detained democracy activist Wilkins LaGuerre, "We were like animals. When they put you in the middle of a desert with barbed wire around, you don't look like a human being any more."[28] In 1991, a *New York Times* article quoted Jean as describing the sanitation problems at the camp, with "latrines . . . brimming over," water that "gave you diarrhea," and "rats [that] crawled over us at night." This puts immunocompromised detainees at risk for secondary zoonotic infections transmitted between species. Jean spoke as a collective first-person refugee subject who, upon encountering these realities of detention, depicted these interspecies relations that subjected refugees to new microbial vectors (evidenced in diarrheal diseases) as compromising the dream of an anthropomorphized space of dignity: "When we saw all these things, we thought, it's not possible, it can't go on like this. We're humans, just like everyone else."[29]

Jean cites a list of US military failures to shield the prisoners from encounters with dangerous animals—which underpinned the sense of their own animalization—as her personal basis for deciding to organize the hunger strike. As Jean describes in her comments to Paul Farmer, the living area was open to the elements, and the prisoners' bodies were regularly brought into a space of unwanted intimacy with reptiles and insects. In addition to using animal figures to portray the indignity of detention, this would violate modern medical expectations regarding proper sanitation. The unexpected touch of small animals crawling and flying across the space of incarceration disrupts the sense of space and scale for the anthropomorphized body. Reflecting the normative acceptance of the ill treatment of stray dogs and livestock as disposable forms of life, the image of the subjection of Haitian bodies to pestilent insects and lizards is further likened to the confinement and indignity meted out to farmed animals. Above, Jean describes being corralled by military police using tractors. She also mentions not being "able to turn around without being injured by the barbed wire," an observation evoking descriptions of concentrated animal feeding operations and other forms of intensive confinement on factory farms. Another unnamed detainee whose protests brought police retaliation told the *Nation*: "Since we left Haiti last December we've been treated like animals. . . . I was chained, made to sleep on the ground. . . . We were treated like animals, like dogs, not like humans."[30] A Haitian released from the Krome immigration detention center in Florida made a similar statement following the hunger strike against unequal treatment of Haitian and Cuban prisoners there: "This was supposed to be the land of paradise, but I saw the real thing. They treated me like an animal, I can't even say they treated me like a dog because dogs in this country get better treatment than Haitians."[31] Testifying in US District Court, one Guantánamo survivor called the camp a "park for pigs."[32] The camp concentrated imprisoned populations and subjected them to unpredictable scenarios of interspecies touch—a deprivation of shelter that served as a soft form of state torture.[33] The Haitian refugees in turn read this treatment as the state's effort to produce their bodies in a racialized image of dependency, indignity, and excess. At a meeting of the hunger strikers, one refugee suggested, "We know we are black and our color makes us suffer."[34] Empire's precarious rendering of Haitian bodies reinforces the colonial logics that animalize black bodies, figuring Haitians as infected and infectious. The horrors of camp life in these narratives dissolve everyday forms of species containment that underpin notions of domesticity, dignity, and rights central to liberal humanitarian logics. This reflects Ranjana Khanna's assertion that a "taint is carried in the concept of dignity." The realities of instrumentalization of life are "inscribed through" the human's "relations to the state, to the conceptions of the international, and to the question of species."[35]

These narratives of camp life contrast deeply with the elaborate humanitarian justifications for the retention of Camp Bulkeley. Presidential candidate Bill Clinton made a campaign promise to end the segregation program, but the Clinton administration later continued the government's legal defense of the HIV camp at Guantánamo. The medical and mainstream presses took for granted the humanitarian intent of the policy, proclaiming without question the state's "compassion" and the ethics of "military physicians" who "worked hard to treat the Haitians."[36] While discouraging migration on poorly equipped boats was an ostensible humanitarian goal, Clinton made clear that US Americans, from his view, could not afford a "mass exodus of refugees" given the estimated 300,000 Aristide supporters in the Louvalas people's movement who remained in hiding following the coup.[37] Indeed, the state's understanding of its supposed responsibility to securitize against the economic and public health threats of mass migration became the major legacy of the litigation over the refoulement of Haitians.[38] The twin specters of invasion (the mass of the dispossessed who could theoretically arrive from the world over) and contagion (the absolute refusal of HIV-tainted bodies) coalesce in the ambivalent figuration of a humanitarian camp. The logic of control, in the words of Judge Sterling Johnson, represented "potential public health risks to the Haitians held there"; Johnson called Camp Bulkeley "nothing more than an HIV prison camp" and a "tragedy of immense proportions."[39] For Paik, the camp "reduced" the refugees "to mere bodies to be minimally managed"—though, I would add, even this fact of detention could be easily disavowed given the state's control of information. Rendering Haitian refugees precarious, the state capitalized on what Talal Asad calls the "combination of cruelty and compassion that sophisticated social institutions enable and encourage."[40] The state was protecting borders, ostensibly, to protect life on both sides; yet the regulation of life in the camp was accomplished through an architecture of deprivation within indefinite detention. The concentration of Haitian refugees as HIV hybrids meant their subjection to an unacknowledged affective warfare in which they baked, in boredom, under the sun, overrun by rodents; soaked, under improvised roofing, among snakes; eating in the open from cans, ingesting the lively surprises of insects taking flight to the mouth; proximate to the barbs of wire and the stings of bees.

The Weaponry of Life Itself

The war over HIV in the Guantánamo prison camp targeted interspecies connections between bodies in the space of incarceration. Struggles between prisoners and military security took aim at the deep structures of refugee bodies, where immune defenses entangled with viruses and other bodies in the environment. Given that the state maintained regular monitoring of prisoner health indicators, the Haitian refugees learned quickly that the condition of their own bodies was a site of concern to the state and thus could be politicized, albeit out of a position of extreme constraint. A mix of malnutrition, immune surveillance, and contact with secondary infections created conditions in which refusing the tools the state used to extend life (and thus justify incarceration) could modulate the presentation of HIV quarantine to the state and broader publics, throwing the camp regime into crisis. A refugee hunger strike publicized criticism of incarceration while accelerating the violences of deprivation in order to unveil the lie of a humanitarian camp.

Resistance to Guantánamo was thus not only a question of legal status and competing discourses on immigration. In fact, because of the severe constraints on speech and movement, incarcerated refugees engaged in warfare that took on the form of life itself, politicizing the inner workings of the body in order to disrupt the emergency measures of the racial state. This was a development that was not limited to the space of incarceration in Cuba. In fact, hunger strikes are common in US immigration jails.[41] And in the rapidly changing political environment of the early months of 1993, public protests and hunger strikes supporting the asylum claims of Haitian political refugees erupted across several sites. As the Guantánamo lawsuit reached the US District Court in New York, an international trade embargo put pressure on the military dictatorship led by General Raoul Cédras and Marc Louis Bazin to restore Aristide to the presidency. Meanwhile, large numbers of refugees continued flowing north and immigration authorities began housing Haitians in local jails across the Gulf Coast in Florida, Louisiana, and Texas. Witnessing the quick release from the Krome immigration jail of forty-eight Cubans who arrived on a hijacked plane from Havana, Haitian detainees at Krome immigration jail in South Florida launched a hunger strike for their own immediate release on December 31, 2002. The strike, involving 158 prisoners at its height, lasted weeks and brought threats and intimidation from staff at the prison, who threatened to force-feed the strikers.[42] Although the INS did not make any concessions in this case, it lent momentum to public demonstrations organized by Haitian activists in Miami connecting the plight of immigrants held in Cuba, Florida, Louisiana, and Texas. By late January 2003, the HIV-positive refugees at Guantánamo would launch their own indefinite hunger strike that included the refusal of medicine. Hunger striking helped bring about the judicial ruling against segregated incarceration for Haitian refugees, but it also likely contributed to the premature deaths of prisoners following release. The rationale for the strike at Guantánamo echoed the one articulated in the letter released earlier in the month from the Florida prisoners, whose asylum claims had stalled for months: "Now we want our freedom without conditions. . . . Freedom or death for Haitian refugees in Krome."[43]

Surveillance of antigen and immune cell counts in refugees' blood tests was central to the structure of imprisonment in the camp, and became the legal basis for releasing Haitians once the Haitian Centers Council (a coalition of Haitian civil organizations) filed a lawsuit on behalf of the refugees that reached US District Court. In the practice of segregation of HIV-positive Haitians in Camp Bulkeley, seropositivity (the presence of HIV antigens in blood serum) provided the medical basis for quarantine of prisoners from the other "screened-in" refugees. These tests are imperfect since there is a weeks-long lag time for seroconversion, and immigrant advocacy groups also reported a high number of false positives early on at Guantánamo.[44] After determination of an HIV-positive status, the prisoners were regularly checked in a second test for counts of particular white blood cells (CD4 + T helper cells) to keep tabs on the progression of AIDS. If T cell levels dropped below 13 percent or 200 per microliter, a prisoner met the CDC's diagnostic criteria for full-blown AIDS, which was the standard that the district court eventually accepted as the baseline for the proper care mandated by Eighth Amendment protections for the incarcerated. Yolande Jean, with a reported T cell count of 235 (13 percent) in her initial screenings, sat at the threshold of full-blown AIDS prior to the hunger strike. A single infectious disease doctor in the camp clinic (supported by one family physician, five nurses, and twenty-four military corpsmen) carried out these tests, and reported inadequate resources to treat patients with AIDS. The segregated clinic as well as the main base hospital lacked CAT scanners and specialists (in ophthalmology, neurology, nephrology, pulmonology, and oncology, for example) to address common opportunistic infections, including eye infections, meningitis, pneumonia, and Kaposi's sarcoma, a viral-induced cancer impacting 10 percent of AIDS patients. Opportunistic infections signaled the immune system's inability to control the balance of organisms crossing the space of the body, and a diverse population of microbial forms—from fungal spores that threatened meningitis to infectious bacteria to amoebas and other parasites arising from poor sanitation—threatened to further undermine refugee health. Tuberculosis was the most prominent of these opportunistic infections, and the camp clinic treated several HIV-positive prisoners with supervised TB regimens.[45] The biomedical problem of opportunistic infections related to a legal problem, since the basis of liberal rights lies in a subject contained within a self-protecting immune body. An autoimmune disorder signals the inability of that body to self-recognize and recode and thus to produce itself out of the immanent interspecies contacts that constitute the environment of every vertebrate body.[46]

Given the unique stresses of the migration experience and camp life, Haitian refugees were already at risk for secondary health problems, and spoke of diarrheal diseases, which contribute to the common problems among AIDS patients of anemia and malnutrition. Because HIV is a rapidly mutating virus that most directly attacks the human immune system, each case takes on specific characteristics, deeply impacted by the use of medicine, health indicators, and secondary infections. One of the most devastating facts of incarceration at the camp was that poor sanitary conditions threatened to make the onset of AIDS more rapid among those with no outward signs of the syndrome, since it is often secondary infections that bring an end to the latency period of HIV and allow the virus to begin a full-body assault on immune cells.

After offering lengthy individual testimony, protesting directly against their captors, enlisting the help of outside lawyers and activists, and witnessing the broken promises of the Clinton administration, quarantined refugees recognized that increasing their vulnerability by hunger striking was one remaining option for resistance to incarceration at Guantánamo. Already figured as human-viral hybrids through legal isolation and medical surveillance, refugees exercised a constrained choice to live in the open, among the camp animals. They refused state provision, opening the body not only to hunger and the elements, but also to the viral, fungal, and bacterial transits that threaten the immune capacities. The prisoners initiated the hunger strike on January 29, 1993, one and one-half years after the military began segregating HIV-positive refugees. At this point, the legal case brought by Yale lawyers and the Haitian Centers Council appeared in the court of Judge Sterling Johnson of the Eastern District of New York. Lawyers were also negotiating with the Clinton administration to bring about the promised closure of the camp. Increasingly skeptical of legal and political processes that kept their bodies and voices at a distance, hunger striking allowed the refugees to establish an alternative critique of detention that circulated in the media, the medical establishment, and the courts. Prisoners not only proclaimed freedom of the group to outweigh their individual deaths, but also attempted to mobilize their own illness toward multiple strategic goals.

Newspaper accounts of the strike described forty "limp Haitian refugees" lying on their backs atop a stage on the soccer field, surrounded by razor wire.[47] Hunger striking in this case did help the refugees gain a certain type of visibility, but a visibility that was at first complicated by the media's predictable sentimental renderings of victimhood. Hunger strikes offer a generic form to specific protests, putting pressure on liberal-humanist state systems that claim to defend life and liberty. For this reason, hunger strikes have often been strategic organizing techniques in decolonial struggles.[48] While the Haitian case is similar to many other examples of politicized self-starvation, the Haitian detainees were already physically compromised by disease and displacement, and were not in a position similar to anticolonial rebels and political leaders for whom engagement was a matter of individualized, heroic commitment to a movement. Haitian refugees had survived such a liberation struggle at home, but when displaced to Guantánamo had been forced into unfamiliar circumstances (including, to some, an unbelievable HIV diagnosis) that gave a unique shape to their activism. The hunger strike takes on a particular cast when it interrupts the narrative that quarantine is humanitarian provision.

Transcending mere publicity, hunger striking transforms the conditions of life. While the hunger strike wagers individual life on the common cause, it should not be understood only as a transaction of biological deprivation for political power. Extended hunger striking does real work on the form of bodies, staging deprivation as an extreme test of the threshold separating life and death, body and spirit. This makes a spectacle of both the queer transition of the deprived body and the highly gendered ascetic power associated with extreme bodily control. As the most vocal of hunger strikers, Yolande Jean told a reporter, "Let me kill myself so my brothers and sisters can live. . . . Mr. Clinton must not know what is inside the Haitians to want to do this. If it is martyrs he wants, I will be glad to be the martyr so everyone else can leave."[49] The interplay of suffering and control is the basis of a logic of martyrdom that makes embodiment itself into a site of solidarity. It ritualizes community ethical response to the vulnerable body as it intensifies time pressure on political process—the state must respond or there will be death.[50]

Hunger strikes can be read as cosmopolitan invocations of humanity; they can also be interpreted as entering into an ethical relation of sacrifice (often based in an economy of spirit in which the self exceeds biological life), which is distinguished from the related option of a self-annihilating suicide. While the Haitian refugees, who consistently appealed to notions of a common humanity, avoided any mention of cultural difference and generally avoided couching their struggle in religious language, their responses to reporters' and lawyers' questions about the hunger strike were suffused with the Haitians' situated histories of political protest that integrated political doctrine with spiritual force.[51] Elma Verdieu distinguished death by hunger striking from suicide, arguing, "God will receive my soul" for the form of communal, ethical sacrifice.[52] When asked to explain the strike in the context of an INS deposition, Jean explained, "We started the hunger strike so that this body could get spoiled and then the soul can go to God."[53] Such an explanation both draws on the logic of martyrdom and makes a clear strategic demand to the INS lawyer: free us or bury us. Hunger striking makes a spectacle of the deathly architecture of camp deprivation, demanding that the state that operates the system demonstrate its publicly avowed commitment to life. It also exceeds the anthropocentric domain of a liberal social contract, invoking a spiritual contest to maintain selfhood apart from the biologized body at risk of death.

While on the first day of the strike most of the adult camp residents participated, the number later stabilized with a core group of up to forty strikers, mostly women. At least twelve of these individuals experienced fainting from exhaustion, and Yolande Jean fainted during her deposition with an INS attorney. As the strike progressed and a handful of strikers continued to refuse food, antiretroviral medicine, and nutritious liquids, they were literally accelerating the destruction of their immune systems. Paik quotes a number of depositions that argue that Guantánamo "is not a place for human beings to take medication."[54] The refugees refused treatment and argued that their HIV diagnoses (which had often been announced to them by INS officials rather than doctors) were fraudulent, promising to consult independent doctors upon release.

Mainstream media omitted descriptions of the assaults and threats that military police, in full riot gear, leveled at the protesters. During the district court trial, the plaintiffs forced the INS to disclose a video that was played before the judge, graphically showing a nighttime raid on sleeping hunger strikers by an estimated 300–400 military police, who assaulted, restrained, and intimidated startled refugees (including children) as a fighter jet flew overhead.[55] At the beginning of the strike, lawyers for the Haitians had tried more modestly to convince the Clinton administration that release was "reasonable and cost effective"; the hunger strike came as a surprise (even an annoyance) to them because it insisted on immediate justice.[56] Lawyers had to rework their strategy and broadly contest both refoulement and HIV exclusion as a matter of law, a crime of the state, rather than an expedient. This more aggressive constitutional strategy, forced by the moral claims and spectacle of striking ill bodies, yielded much faster results that might never have been obtained otherwise. Recognizing the deterioration of the health of several strikers, Judge Sterling Johnson in March 1993 quickly issued an interim order "to prevent any loss of life" by requiring a higher standard of medical care for prisoners with T cell counts under 200 per microliter.[57] Although the refugees were wary of further medical testing, their eventual acceptance of these tests led to the release within ten days of dozens of the sickest refugees. The time pressure of the strike strategy forced the judge to codify in law the threshold of unacceptable bodily vulnerability, which proliferated new types of state surveillance of the HIV-positive immune system. This was the first time a legal standard of incarceration hinged on state surveillance of the deep structures of HIV-positive bodies with reference to the CD4 or T helper cell count. Tragically, the lived consequences of the strike became clear with the death of one of the strikers within three weeks of his move to Miami. Johnson eventually upheld all of the plaintiffs' arguments in a sweeping judgment in June 1993, at which time the Clinton administration reluctantly announced a plan to remove the rest of the HIV-positive Haitians to the mainland United States. The ruling presented the first legal rejection of the unfettered exercise of domination over Haitian refugees at Guantánamo. The hunger strike was a source of power and solidarity, but also a measure of desperation; at least four refugees attempted suicide in the last year of detention.

Following the slow release of the remaining Guantánamo detainees, ending in mid-1994, death rates were high among the formerly incarcerated HIV-positive refugees. An AIDS activist in New York in 2003 estimated that up to half of the prisoners had died, and that some of these deaths were accelerated by the conditions of detention.[58] The affective warfare carried out by both sides on Guantánamo had life-and-death consequences for HIV-positive refugees, even though these consequences were not reported widely in the media once the state released them to the continental United States. It was the hunger strike, with its ritualized spectacle of deprivation and intensification of interspecies contact, that allowed for a quickening of the politicized contest over refugee bodies. With the release of refugees from imprisonment, there was less incentive for media reporting and a release of pressure on the state to act. Worse, the Supreme Court later overturned Sterling's decision and allowed the Coast Guard to restart refoulement. Over a decade later, most of the asylum claims still had not been processed, nor had the noncitizen status of children born on Guantánamo been resolved.[59]

Coda: The Repeating Camp since 9/11

In this chapter, I have argued that the violence of the Haitian HIV camp was carried out through practices of concentration, deprivation, and regulated transspecies intimacy. As reflected in the testimonies of quarantined Haitian refugees, the resulting hunger strike and the eventual form of the legal process for HIV-positive Haitian refugees centered on the politicization of the deep biological structures of embodiment at the so-called humanitarian camp. Today's Guantánamo hunger strikes among war prisoners from Africa, the Gulf states, and South Asia, who in most cases face no formal charges, have also occurred under harsh conditions, bringing those released to commonly compare the prison to a zoo, as did the Haitians years earlier.

Despite the significant differences between quarantined Haitians in the 1990s and prisoners of war today, questions of humanitarian application of law and care abound at Guantánamo. Although the emergency detention of so-called enemy combatants is based on legal reference to the detention of Nazis following World War II, the Justice Department has also used the same justification for the suspension of habeas corpus as it did in the Haiti case, arguing that domestic law does not apply in Cuba. And even as the courts have refused this argument, the executive branch has been able to continue indefinite detention with the help of a compliant appeals court in Washington, DC, that consistently grants executive privilege. Meanwhile, military officials effectively censored media representation of the hunger strikes that began in 2005, suppressing or delaying information on causes of prisoner deaths and the numbers of prisoners engaged in strikes. This situation continued until a mass hunger strike at Guantánamo began to make headlines in 2013, forcing the Obama administration to restart efforts to relocate prisoners. Notably, the commander of the detention camp at Guantánamo has defined hunger striking as a form of "asymmetrical warfare," lumping it in with suicide bombing and guerrilla warfare. In the meantime, the military has produced manuals enumerating standard operating procedures in the face of emergency detention and hunger striking. These documents acknowledge the strategic psychological value of internment in maintaining US international dominance and the significance of hunger striking as a form of resistance that needs to be managed through standard techniques.[60]

As in the case of the Haitian prisoners, today's Guantánamo features regular hunger strikes, suicide attempts, and a response by the state that attempts to advertise its conformity to standards of humane care. Still, the language of shared humanity, martyrdom, and spiritual struggle at the border of life and death, human and animal, appears regularly in today's Guantánamo prison writings as it did with Haitian prisoners. The use of spiritual rhetoric in prisoner writings, testimony, and interviews should not be read as an expression of culture but as a method of extending the agency of resistance beyond the anthropomorphized limits of body and life. Prisoner #261, Jumah al-Dossari, wrote a poem that spectrally offered up his own body to surveillance as evidence of the crimes of the racial state:

Take my blood.
Take my death shroud and
The remnants of my body.
Take photographs of my corpse at the grave, lonely.
Send them to the world,

To the judges and

To the people of conscience.

"Death Poem" speaks to the brutality of incarceration and the removal of Guantánamo and secret CIA prisons from public visibility. Under heavy censorship, only a handful of such poems made it beyond the barbed wire fences, as the Department of Defense claims that "poetry . . . presents a special risk, and DoD standards are not to approve the release of any poetry in its original form or language."[61] Al-Dossari was released from Guantánamo in July 2007 after a reported thirteen suicide attempts and after disseminating, through his lawyer, some of the most widely cited detainee testimony regarding prisoner abuse at Guantánamo. Al-Dossari's writings capture the force of imperial power upon his own body at the site of detention: "How I wish my memories and my thoughts could be forgotten. But for me, in forgetting it and its effects, there are still memories, lifelong evidence of what happened to me in my wounds, my afflictions, my pain and my sadness."[62]

One hunger strike began in June 2005, when over forty prisoners refused food across all five camps at the center. The strike ended after twenty-six days, when prison officials claimed they would enforce the Geneva Conventions. This promise was quickly broken, and a series of hunger strikes ensued to continue protesting the conditions of the camp, most importantly violent abuses of the prisoners.[63] Those prisoners who have since maintained the strikes the longest — and with success at gaining some concessions from prison officials — suffered health problems and were taken to the camp hospital to ensure they did not die. The first standard operating procedure established for the hospital was for force-feeding striking prisoners, who are strapped into a special chair for hours at a time to ensure they keep down the tube-fed nourishment. Al-Dossari had to receive a blood transfusion after one hunger strike, and Adnan Latif, a prisoner who died in 2010, engaged in hunger striking and threatened suicide in a letter that recalls Yolande Jean's desperate death letter. Latif denounces a "prison that does not know humanity," that "does not differentiate between a criminal and the innocent," claiming, "I will not allow anymore of this and I will end it. . . . With all my pains, I say goodbye to you and the cry of death should be enough for you." Critically, Latif distinguishes between "the imposed death" that he already lives in confinement and the limited "freedom" in which he "choose[s]" his "own end."[64]

To date, nine prisoners have died at the post-9/11 camps at Guantánamo.[65] Yet the forms of depersonalization at Guantánamo must not be read as excep-

Figure 5.1 US Department of Defense promotional image of the Qur'an at Camp Delta, Guantánamo Bay Naval Base. Photograph by Army Sergeant Sara Wood, April 5, 2006.

tional; even the Krome immigration jail had prepared to force-feed hunger strikers. Although Guantánamo has been the site of discriminatory AIDS imprisonment and the persistent denial of access to counsel and other basic legal rights, it is actually similar to some US prisons (especially maximum-security prisons that rely increasingly on solitary confinement and force-feedings) in its violent wrenching of the personality of detainees through techniques of isolation, sensory deprivation, and denial of rights to expression and religion.[66] Yet Guantánamo officials had elaborate public relations plans to present the camp as a modern facility rather than an extrajudicial space of torture and incarceration. They organized staged press tours placing emphasis on the provisions, recreation opportunities, and medical facilities. As in the case of the Haitian refugees, the government became well versed in publicizing the cultural differences that it proclaimed it respected (despite prisoner claims of regular desecration of the Qur'an, denigration of Islam during interrogations, and the ominous broadcasting of the call to prayer from the guard towers). One image from these press tours in particular struck me. In it there hangs a blue and gold Qur'an, suspended from the wire fencing of the cell, held by a white surgical mask (figure 5.1). Attempting to demonstrate religious sensitivity, officials advertised their aid in helping inmates properly store the holy book away from contact with the floor. The surgical mask—perhaps the most common symbol of hygiene in allopathic medicine—is here deployed to maintain the purity of the Qur'an. Medical and health technologies, then, apparently serve as agents of incorporation, helping display the possibility of inclusion of religious/cultural diversity in disciplinary techniques at the site of incarceration. These practices are effective within the realm of public relations precisely because, being associated with progress and the health of the population, they can become indices of care and humanitarian relief. Health and security, humanitarianism and confinement, emerge in tandem at Guantánamo.

Species War and the
Planetary Horizon of Security

Infectious disease is one of the great tragedies of living things—the struggle for existence between different forms of life. Man sees it from his own prejudiced point of
view; but clams, oysters, insects, fish, flowers, tobacco, potatoes, tomatoes, fruit,
shrubs, trees, have their own varieties of smallpox, measles, cancer, or tuberculosis.
Incessantly, the pitiless war goes on, without quarter or armistice—a nationalism of
species against species.
—HANS ZINSSER, *Rats, Lice, and History*, 1935

The annihilation of certain species is indeed in progress, but it is occurring through
the organization and exploitation of an artificial, infernal, virtually interminable survival.
—JACQUES DERRIDA, *The Animal That Therefore I Am*, 1997

Despite German American physician Hans Zinsser's depiction of infectious
disease as a sign of nature's timeless and intractable "nationalism of species
against species," replete with tragic but inevitable consequences of death and
debility, I have argued in this book that disease interventions are political processes, ones that link media, bodies, institutions, and medical technologies
in order to securitize privileged forms of circulation against the queer potentials of interspecies contact. Such containment has been a consistent preroga-
tive of the US security state throughout its twentieth-century experiments in
trade, military, and territorial expansion, crossing a variety of institutional
projects to control space, engineer immunity, and enhance the scale of intervention against infectious disease. Species are not, then, simply in protracted,
ancient struggles for survival. If—in today's era of mass extinction and ecological crisis—critical social and political theories are beginning to imagine

the intensification of species war coincident with current forms of empire, this would not simply be a war between independent species struggling for evolutionary advantage. The government of species has a history, one that suggests that power takes shape in part through the organization of bodily forms and relations.

Contemporary modes of violence and extermination may indeed, then, be intimately linked to Jacques Derrida's vision of an "infernal, virtually interminable survival."[1] Derrida's words relate to the specific context of industrial animal agriculture, where the grand scales of mass killing at the slaughterhouse occur not in the service of simple extermination, but in the persistent rendering of life into capital.[2] This idea has bearing not only upon the practices of the government of species described in this book, but also on the more general understanding of contemporary neoliberalism and its futures. Austerity, inequality, unemployment, and chaotic biological and economic risk — the symptoms of systems conducting greater masses of bodies and species through the processes of slow death — signal the exhaustion of populations under the combined banners of free markets and security; the ever-narrowing securitized zone of human life regenerates against the backdrop of a collapse of environmental and social systems, producing a kind of "infernal, virtually interminable survival."

Yet one difficulty of describing these connections through an anthropomorphic account of the political — one that focuses on human society to the exclusion of its embeddedness in ecological systems crossing species — is that we may miss how trends toward austerity and a shrinking state coincide with the neoconservative expansion of security interventions across the domains of life and planet. Thus it is necessary to explore the potential for empire's supplanting of an earlier phase of neoliberalism with emergent forms of securitization. The concept of the government of species as outlined in this book offers a genealogy of the shift toward the securitization and remaking of biological life coincident with privatization, the dismantling of state investments in social welfare, and imperialist efforts to contain the transnational effects of decolonization. In the pages that follow, I offer some concluding thoughts on how this book's analysis of the conjunction of biopower and necropower contributes to understanding these political trends. This is especially important as today free-market trade imperialism opens into full-spectrum military-industrial defense, an attempt to securitize posthuman environmental, social, biological, and technological assemblages.

Noting that the early neoliberal "Washington Consensus" advocating free trade, austerity, and privatization seems to have collapsed internationally despite those cases—like Greece—where it can be forcibly imposed, I reflect in this epilogue on some central questions concerning the futures of neoliberalism and security. In particular, I explore how Left critiques within the academic fields of postcolonial studies and American studies can respond to (1) the difficulty in capturing the persistent materiality of race and other forms of difference as they relate to debates over neoliberal forms of economic and ecological crisis; (2) the question of whether emerging posthumanist conceptions of risk reflect neoliberal logics or their transformation into a new post-neoliberal phase of security; and (3) confusion over how securitizing logics affect the political form and representation of US empire given the mainstream political adoption of the language of imperialism. These questions bear further on how to think about the interspecies history of US empire in relation to changes in the global economy and the rise of environmental politics and powerful "semiperipheral" or "first-third-world" states.

I have argued throughout this book that race has long played a central role in the US security state, particularly as it works to map and aggregate biological risk. From the depiction of native Hawaiians as vanishing victims of Hansen's disease in the late 1800s to the imagined extermination of North American populations by epidemic smallpox in the 2003 Iraq invasion, many public visions of infectious disease outbreaks figure particular social groups as conduits for the aggression of microbial species, the incessant processes of viral and bacterial emergence that always threaten to disrupt the fabric of life itself. Dread life processes the living potentials and fears of contact into specific architectures of power, exercised in the medical incorporation of racialized groups, the expansion of markets and institutions, and the configurations of species and distribution of technologies across dimensions of time, space, and scale. Dread life's aesthetic and ontological linkage of race and species continues to play a persistent role in the work of the security state precisely because it offers road maps for navigating the uncertainty of contact internal and external to the body. Race is a structured element of the government of species, not because racialized groups bear unchanging relations to the state, but because racial form generates national sentiments and strategies of institutional incorporation—ways to flexibly aggregate vital potentials into forms of intervention when there is otherwise only a generic sense of vulnerability or uncertainty. I am describing race as a set of technologies connecting feeling to representation and giving form to insecurity rather than only a content-driven set of policed phenotypes. Race is fluid, emergent in the crises of interspecies contact that generate politics. Despite the late twentieth-century myth of a postracial United States, the security state and the private institutions upon which it relies continue to depend on various forms of racial knowledge and profiling to justify funding and legal authority.

While in this sense neoliberal and neoconservative logics converge in the nexus of state and corporate efforts to control life, the question remains as to whether this nexus represents an emergent order in which human security supplants austerity.[3] The racialization of uncertainty and risk in disease control offers some broad lessons about US and international health delivery, ones that scramble certain accounts of neoliberalism's homogenizing logic. Today life can be regulated intricately against a specter of insecurity that increasingly recedes into a vision of aleatory risk: an emergent "environmentality."[4] The layering of space, time, scale, and species is furthermore geared toward environmentally targeted ideals of human security increasingly adopted in the state's forms of risk assessment.[5]

The passage in the United States of the Affordable Care Act might seem to contradict this potential turn from a laissez-faire neoliberalism to an interventionist human security, as it depends on privatized insurance markets. Yet even if it follows logics of privatization, it is also suffused with emergent logics of human securitization against a specter of environmental risk; these logics allow for different types of actors, ideologies, moral claims, and epistemologies of risk to come together in processes of securitization. Witness a key moment in the oral arguments over the bill, where Supreme Court justice Stephen Breyer compares the universal insurance mandate to the vaccination authority highlighted over a century earlier in *Jacobson v. Massachusetts*. Breyer explains the new US mandate for individuals to purchase private care with reference to the necessity of enforced vaccination during an epidemic, echoing the exact language that the World Health Organization and US biosecurity officials used to depict the "sweeping" threat of epidemic smallpox in the lead-up to the Iraq invasion: "A disease is sweeping the United States, and 40 million people are susceptible, of whom 10 million will die; can't the Federal Government say all 40 million get inoculation? So here we have a group of 40 million, and 57 percent of those people visit emergency care or other care, which we are paying for. And 22 percent of those pay more than $100,000 for that. And Congress says they are in the midst of this big thing. We just want to rationalize this system they are already in."[6] Although Breyer may simply have wanted to defend the state's authority to create a universal health program, the legislation offered him a logic of risk preparedness through insurance pooling. Combining a long-standing rhetoric of species war with the statistical aggregation of population risks, Breyer's numbers turn toward mass catastrophe to justify intervention. The statistical sublime of this sweeping epidemic underwrites both a turn toward the privatization of risk and an attempt to incorporate futurity and unpredictability into the sphere of human security. This logic articulates the connection of neoliberal austerity with neoconservative preparedness, couched in the language of species war innovated by Justice Harlan in the *Jacobson* case a century earlier.

This form of preparedness has a connection to the problem of race that I have explored throughout the book. Liberals in the United States have logically responded to the situation of deepening inequality domestically and internationally through an embrace of the social welfare state, imagined as the real defense against various forms of proliferating precarity intensified by neoliberalism. But what if this response erroneously characterizes the form of power emerging in the nexus of austerity and security? The emergent logics of human security reflect how today even social reform policies are worked through a fear of racialized dependency. It is important to remember that the strengths of social welfare were challenged early on in the United States and Britain precisely at the moment when immigration reforms led to a racial backlash aimed at the supposed parasitic dependency of diasporic populations and the integrationist state itself. As I explained in chapter 4, neoconservatives turned the idea of the hubris of government's attempts to contain a complicated world of race relations (domestically) and communism and decolonization (internationally) into twin attempts to dismantle the welfare state and to militarize foreign policy. Thus the turn from a social welfare state to both austerity and security could be interpreted as a response to the apparent loss of racial privileges coeval with the mass arrival of third-world immigrants. Even as protecting and expanding social policies represents an important and necessary response, such agendas will face at least two difficulties. First, they will be increasingly vulnerable to austerity and to panics over migration that have grown stronger since the recent interventions in West Asia have created mass refugee crises. Second, as securitizing logics do expand rationales for human security, the emphasis in social programs is more likely to be oriented toward crisis thinking, transforming the logics of welfare from within. Social programs are more likely to be governed by logics of systemic risk and preemption rather than social provision or socioeconomic equality. They may furthermore be captured by discourses of civilizational or cultural difference that distribute social goods according to a different set of criteria than in the past.

The racial and colonial legacies of medicine and public health compound these uncertainties over the future of social democracy in the international context, particularly as it relates to global public health. The intensifying problem of antibiotic resistance has been accelerated by a pharmaceutical empire that has unequally distributed antibiotics across the international division of labor and has improperly allowed industrial agricultural use of the drugs. At the same time, many publics openly reject or question medical and sanitary authority backed by Western nongovernmental organizations and international organizations. In the introduction, I explained that a technique like vaccination had become deeply politicized, with Donald Rumsfeld using it to advertise the nation's shared vulnerability to Saddam Hussein, and the CIA using it as a clandestine weapon in the hunt for Osama bin Laden. In the twenty-first century, such nefarious uses of health internationalism for strategic military or political objectives—combined with many failed international medical and sanitary projects in the Global South—continue to threaten the advertised neutrality of health and medical interventions. This is a problem that endlessly frustrates global health experts who hope to battle "superstitious" distrust of modern medical knowledge and practice across poor countries and, increasingly, in the United States and Europe as well. As I explained in the discussion of paranormal discourse in Puerto Rico in chapter 3, it is necessary to understand such political logics of suspicion. Resistance to the racialized differences in health and medical outcomes is political, not simply cultural. It is alive and well today, apparent in the Taliban's execution of health workers in Pakistan; the mistrust over sexual education programs in southern Africa and Ebola screening in West Africa; and the rising antivaccination campaigns in the United States and Britain, to name a few examples. As I have argued throughout the book, disease interventions are constantly confronted by the challenges of patient activists, related social and political movements, and the unruliness of bodies and media assemblages. The results are uneven. For example, in attempts by patients to gain access to experimental Hansen's disease treatments in chapter 1, the state was criticized as the barrier to obtaining more expansive care. This dynamic was reversed in chapter 5, where racialized HIV quarantine involved a violence that completely broke down trust in doctors and their diagnoses.

My point in this book has not been to denigrate medical sciences and public health as inherently violent or exploitative, nor to idealize patient activists, writers, or research animals as sites of resistance to the imperial order. Although Frantz Fanon once emphasized the Manichean resistance of the colonized body to dissection by the disciplines of medical surveillance, the various cases of disease intervention I have reviewed demonstrate more complex biopolitical relations, especially once patients actively take part in treatments and when medicine links human and animal research subjects across circuitous geographies of pharmaceutical intervention. Health and medicine are embedded in logics of intervention that are not internal to their own practice. One lesson of this book, then, is that such disease interventions have never followed a strictly scientific logic guiding which diseases to target and how to produce therapies and policies to maximize public good for humans writ large or even for specific national communities. They are embedded in dominant forms of state vision and media assemblages that take planetary space and interspecies relation as proper sites of intervention and expansion, even as they also proliferate racialized visions of freedom and precarity, debility and independence. Thus more equally distributing health care is undoubtedly one important political strategy, but it may also open new forms of imperial securitization that layer power across an expanded range of bodies and environments.

Foucault suggests that at the very moment that security apparatuses animate the state and the human as the privileged domains of sovereignty, they simultaneously unveil the state as unable to control life itself, as being dependent on circulatory processes that move laterally beyond its explicit exercises of power. This is one reason that it is necessary to carefully review the spectacular visions of environmental crisis that guide some current trends in the environmental humanities, which have been especially popular in postcolonial studies. These studies have at times turned to the environment and planet in ways that shore up a concept of universal human sovereignty, a trend that may be more a reflection of empire's logics than a critique of them. One prominent critic has argued for a universalist species thinking given his assessment that the fossil fuel economy makes the human itself into the planet's primary historical-geophysical agent.[7] Yet in the last decade, these forms of planetary feeling—which universalize the sense of being out of joint with the evolutionary and geological time scales of the planet—converged with imperial attempts to more broadly technologize inner and outer spaces, to engineer empire's forms of human security against intensifying precarities. Mary Louise Pratt helpfully attempts to situate this public sense of the uncanny separation of human time from planetary time as an outgrowth of capitalist generation of risk sociality and slow death. Pratt, explaining the rise of millenarian visions of mass destruction, quotes anthropologist James Ferguson to explain the emergence of

> vast zones of exclusion inhabited by millions of socially organized people who are and know themselves to be utterly dispensable to the global order of production and consumption. All over the planet, then, large sectors of organized humanity live conscious of their redundancy to a global economic order which is able to make them aware of its existence, and their superfluity. People recognize themselves as expelled from the narratives of futurity the order offers, with little hope of entering or re-entering. This expulsion from history has been accompanied by rapid pauperization, ecological devastation, and a destruction of lifeways unprecedented in human history.... More and more people across the globe speak of the place where they live as a place where there is nothing, where nothing happens, where you would never want to live. Everything of value is elsewhere.[8]

As the idealism of the global is transcended by figurations of planetary crises of human security in the form of extinctions, resource depletion, climate change, and epidemic disease, postcolonial critique is confronted with new holisms (the human as species, the Anthropocene, planetarity) that risk masking the precarious grip of empire's reproduction of contained forms of life. The fact that empire's risks intensify preexisting forms of inequality and migrate outside of the spectacular time of crisis does indeed suggest the need for new forms of representation and critique; however, this means more intricate accounts of bio-necro collaboration rather than new universals that misguidedly assert a generic human domination of the planet.

This turn to planet relates furthermore to some broader shifts in postcolonial studies, particularly its attempt to transcend its traditional divisions of metropole and periphery, and to account for a US empire that appears to present a more diversified geography of political forms, not least in its current misadventures in West Asia. It has now been nearly twenty years since editors Amy Kaplan and Donald Pease published a postnationalist reappraisal of the US empire in *The Cultures of United States Imperialism*, signaling the rise of a new internationalism in a field that had historically celebrated US exceptionalism. Then, it was imperative for Kaplan to chart the politics of the "absence of empire in the study of American culture."[9] Today, the emphatic, axiomatic presence of US empire in a new American studies is matched by overt recognition by state functionaries who debate and idealize the idea of American empire. "We are now an empire," says Bush advisor Karl Rove of the Iraq War, against the protest of Supreme Court justice Antonin Scalia, who nonetheless upheld the imperialist policy denying petitioners legal standing at Guantánamo Bay. In a move evoking Rudyard Kipling's 1899 poem "White Man's Burden," which passed the torch of empire from Kipling's Britain to the United States during its conquest of the Philippines, British historian and Harvard professor Niall Ferguson interprets the United States as taking on the British empire's avowed interest in globally spreading democratic liberties.[10] Such congratulatory invocations of empire portend more complex realities: to what extent does the acknowledgment of US empire work to confirm the exceptionalist logic of US hegemony? Can the neoconservatives' late embrace of empire itself be seen as a form of force projection, rather than a sober assessment of a disavowed historical truth? There is furthermore talk of the rise of China, of the ascendance of the first-third-world states of India, Brazil, and South Africa, and the rising economic and political muscle of fossil fuel–producing states such as Russia, the Gulf states, and Venezuela. Is the will to represent US empire a sign of its very demise?

On Left and Right, answers to these questions have persistently psychologized the imperial project, turning away from analyses that would situate empire in vital and capital processes, in materialisms new and old. For Niall Ferguson, quoting Freud, there is the irrational "denial" of US empire by the American public, as well as the "attention deficit disorder" of politicians (with the exception of the neoconservatives). It is only in the continued and explicit commitment to an evangelism of liberal democracy—including wars in its name—that the United States will maintain its proper global role.[11] In an otherwise sharp critique of the imperial politics of incarceration in the war on terror, one critic has characterized the United States as a "paranoid empire," an empire constituted by a "doubleness with respect to power," moving "precariously between deliriums of grandeur and nightmares of perpetual threat."[12] This assessment helps to capture the doubleness of power I have described as central to dread life, which channels inflationary fears of contact into the force of intervention. However, the psychological metaphor of paranoia constructs an empire constituted by a lack, an "enemy deficit" that must insistently be produced as structural effect of what essentially appears as a psychodynamic process. Failure to recognize and illuminate such disavowals constitutes, for another critic, a practice of history that is "disabled."[13] Given the central role of historicist studies in the post-9/11 work on US empire, it is not uncommon to hear of the public's amnesia concerning the long history of US intervention-ism; indeed, a Google Scholar search for "historical amnesia US empire" re-veals dozens of relevant scholarly studies. Yet rather than psychopathologizing the imperial subject or the US nationalist historian, in the process invoking the moral taint of disability to convey empire's failure of self-recognition, could we not instead track how debates over empire intersect with emergent projects of securitization? How can we disentangle the important work of critical his-torians of empire who have widely detailed the violence of empire from the agendas of neoconservatives who attempt to use such descriptions of power as weapons of force projection within informational warfare?

A limited understanding of empire as the exercise of sovereignty beyond territoriality would overlook security's mutually reinforcing dynamic between territoriality and circulation, a process that has been explored in Marxist ac-counts of the relation between US empire and the development of global capi-talism. Giovanni Arrighi sees this relation between territoriality and circula-tion as reaching its apogee in the unity of settler colonialism, racial slavery, and the rise of the "free enterprise system" in the formation of the United States.[14] Yet Arrighi nonetheless distinguishes this concentration of power from em-pire, arguing that successful US "hegemony" within the world capitalist sys-tem is based on delimiting this territoriality to the continent rather than pur-suing a planetary empire. As I have argued, the delimitation of sovereignty is at times the very basis of territorial expansionism and the tenacity of the accumulation of power and wealth. More recently, Radhika Desai's impor-tant account of the growing constraints on US power in a world of "uneven and combined development" argues that multipolarity is displacing the last gasps of empire expressed in the Bush war in Iraq and constrained by the 2008 financial crisis.[15] Desai gives a compelling account of the limits on US power over the international system. Nonetheless, I see neoconservative logics (such as those of biodefense, environmental government, and human security) as more deeply embedded in institutions than Desai's account allows. Also, in a world of aleatory risk the object of power is no longer humanist containment but posthumanist proliferation: constant intervention and adjustment. It may be that the form of empire no longer depends solely on exercises of sovereign domination but rather on the flexible capacity to intervene. In the nexus of contemporary security, environment, and health, there is continuing evidence of unequal abilities for the United States to intervene and to channel profit through health markets, military technologies, and state formation in various parts of the world. This holds true even if the emergent economic nexus of Gulf oil extraction, US military exports, and Chinese manufacturing link security capacities to decentralizing political economic structures.

Thus more work needs to be done in postcolonial studies and American studies to account for the changing articulations of empire as US dependencies on West Asian oil and South/East Asian manufacturing (as well as a variety of export markets) coincide with shifts in the governing logics of US power. This book has argued that empire persists not only through its accumulation of military force and capital, but also because it can retreat into the apparently neutral domain of life to flexibly regenerate forms of expansion. The disavowal of empire is also in part an effect of the fact of empire's lateral imbrication in vital processes of regeneration and slow death that escape the spectacular temporality of the dramas of sovereignty and resistance. Thus it is especially important to track imperial form, to recognize the effects of material circulation in the emerging constitution of security. As I explained in the preface, life gives form to the political, including forms of imperial domination that transfer affective and institutional energies into the reproductive disciplines of the body. When empire loses its sole character as the sovereign and takes on a circulatory form, it produces complex geographies of contact that embed the regenerative powers of bodies in intricate and intimate forms of warfare.

NOTES

Preface

1. Scott, *Seeing Like a State*.

2. Stoler, "Intimidations of Empire," 4.

3. Livingston and Puar, "Interspecies," 4.

4. Bennett, *Vibrant Matter*; Haraway, *When Species Meet*, 2–4, 15–19. On biosociality, see Rabinow, "Artificiality and Enlightenment."

5. Karl Marx (*Capital*, 326–27n4) argues for a humanist history that tracks the political-economic forces of mass society through the progressive development of technologies. Unlike Charles Darwin's evolutionary theory, which Marx viewed as an unscientific attempt to divine history from the surface of bodily forms, the history of technology allowed deep access to the social organization of human life underlying the mask of ideology. However, the rhetoric of this polemic and its binary logic dividing human and natural histories disavows what Marx performs throughout his chapter in the first volume of *Capital* titled "Machinery and Modern Industry": an account of how industrial technologies entangle human and inhuman forces. Marx repeatedly recognizes that the energies of the laboring body, soil, draft animals, wind, steam, and coal were harnessed through the technological adaptation of labor processes during industrialization; technologies are thus man-made "organs." Marx tends to flatten the form of inhuman energies into mechanical force for the machinery of capital, which begs the question of how his account of industrialization might have been different if it had considered the ecological "limits to capital" that would become central to Marxist environmentalisms since 1970. Nonetheless, his approach opens, even if briefly, a recognition of the entanglement of the human and its social relations in interspecies ecologies. I thus argue against too strong a distinction of new materialisms from old ones, a distinction that merely seeks to reverse the anthropocentrism of Marx's division of human and natural history without anticipating the dialectical synthesis of those binary terms.

6. See Thrift, *Non-representational Theory*.

7. Berlant, *Cruel Optimism*, 18. Berlant builds on the brief sketches of lateral forms of power/knowledge in Foucault's descriptions of human capital, environmental forces as transversal phenomena, and economics as a science. See Foucault, *The Birth of Biopolitics*, 230–31, 261, 286.

8. Puar, *Terrorist Assemblages*, 35. See also Nast and McIntyre, "Bio(necro)polis"; Ahuja, "Abu Zubaydah and the Caterpillar."

9. Foucault, *The History of Sexuality*, 135–45.

10. Foucault, "The Birth of Social Medicine," 135, 150–51; Foucault, *The Birth of Biopolitics*, 271.

11. Rose, *The Politics of Life Itself*; Esposito, *Bíos*; Cooper, *Life as Surplus*; Sunder Rajan, *Biocapital*.

12. For key postcolonial/feminist analyses of biopower from the 1990s, see Vaughan, *Curing Their Ills*; Stoler, *Race and the Education of Desire*; duBois, *Torture and Truth*. For more recent necropolitical critiques, see Agamben, *Homo Sacer*; Appadurai, *Fear of Small Numbers*; Berlant, "Slow Death"; Chua, *In Pursuit of the Good Life*; Hage, "'Comes a Time We Are All Enthusiasm'"; Mbembe, "Necropolitics"; Rose, "What If the Angel of History Were a Dog?"

13. For a provocative reading of the connections between Foucault's late discussions of neoliberalism (in *The Birth of Biopolitics*), the Iranian revolution (in reporting for the Italian newspaper *Il corriere della serra*), and the ethics of the ancient Greek householder (in *The History of Sexuality*, vol. 2: *The Care of the Self*), see Cooper, "The Law of the Household." On Foucault's "political spirituality," see Afary and Anderson, *Foucault and the Iranian Revolution*, esp. 4–5.

14. Foucault persistently retreated from the domains of the animal and of interspecies life, even when he discussed the shepherd as a figure of "pastoral power" (see Pandian, "Pastoral Power in the Postcolony," 90–91) and when he analyzed oil strikes during the Iranian revolution.

15. Alexander Weheliye's important study *Habeas Viscus* was published shortly before this book went into production. Weheliye offers a black feminist critique of the racialized management of the human/nonhuman binary, noting the undertheorization of race in "bare life and biopolitical discourse." While I am unable to substantively respond to his study here, my own approach differs by distinguishing critical theories of bare life from those of biopower, allowing an appraisal of Foucault's unfinished turn toward race and empire in the late lectures as a particular response to neoliberal shifts in the governmental logics of capitalism, security, and militarism.

16. Lowe, "The Intimacies of Four Continents." For foundational postcolonial critiques of the human, see, most notably, Spivak, *A Critique of Postcolonial Reason*; Mignolo, *Local Histories / Global Designs*; Chakrabarty, *Provincializing Europe*.

17. For feminist studies, see, notably, Haraway, "A Cyborg Manifesto"; Schiebinger, *Nature's Body*; Plumwood, *Feminism and the Mastery of Nature*; Butler, *Bodies That Matter*; Grosz, *Volatile Bodies*; Ritvo, *The Platypus and the Mermaid*.

18. Goldberg, *The Racial State*.

19. Cooper, *Life as Surplus*. Cooper's notion of a vital fix draws on Marxist theories of spatial fixes to the crises of capital. See Harvey, *Spaces of Capital*, 284–311.

20. Shukin, *Animal Capital*, 11.

21. Wolfe, *Critical Environments*.

22. Silva, *Toward a Global Idea of Race*, esp. 45–48.

23. Haraway, *Primate Visions*, 10–12. See also Mazel, "American Literary Environmentalism as Domestic Orientalism."

24. Parsons, "Sociological Reflections on the United States," 194; Deleuze and Guattari, *A Thousand Plateaus*. I am informed by the indigenous studies critique of Deleuze and Guattari in Byrd, *The Transit of Empire*, 17–19.

Introduction

1. See Durbach, *Bodily Matters*. The "conscientious objector" entered into English law when parents secured the right to refuse children's vaccinations. The concept was applied to military draft refusal later.

2. See Mariner, Annas, and Glantz, "*Jacobson v. Massachusetts*." In the 1927 case *Buck v. Bell*, Justice Oliver Wendell Holmes cited *Jacobson* to argue that "the principle that sustains compulsory vaccination is broad enough to cover cutting the Fallopian tubes"; forced sterilization was a "lesser sacrifice" than asking "the best citizens for their lives" in war. This justified programs in which 60,000 individuals were coercively sterilized from the 1920s through the 1970s. Although it did not cite *Jacobson* directly, the Supreme Court decision in *Korematsu v. United States* used a similar rhetoric of citizen sacrifice as basis for Japanese internment during World War II.

3. Minor reactions appear in as many as one-third of smallpox vaccination cases, and life-threatening adverse conditions appear in smaller numbers. The Smallpox Vaccination Program delivered nearly 40,000 doses in its civilian program and 625,000 on the military side. US Army Specialist Rachel Lacy died of inflammatory damage to the heart and lungs on April 4, 2003, following vaccinations for several diseases including smallpox and anthrax. Three other deaths due to heart attack and dozens of cases of serious adverse cardiac reactions were reported. See Committee on Smallpox Vaccination Program Implementation, *The Smallpox Vaccination Program*, esp. 46.

4. Donald Rumsfeld, interview with Larry King, *Larry King Live*, December 18, 2002, http://www.defense.gov/transcripts/transcript.aspx?transcriptid=2940.

5. On affective warfare, see Virilio, *Desert Screen*; Massumi, "Fear (the Spectrum Said)."

6. Arrighi, *The Long Twentieth Century*.

7. For "imagined immunities," see Wald, *Contagious*, 51–53.

8. Saldanha, "Reontologizing Race," 20. Saldanha's metaphor of the viscosity of race is useful here in describing the simultaneous fluidity and stickiness of race that allows it to flexibly structure social relations. For additional materialist accounts of race that emphasize its spatial, affective, and phenomenological dimensions (rather than reducing race primarily to a problem of epistemology), see Saldanha, *Psychedelic White*; Puar, *Terrorist Assemblages*; Mawani, "Atmospheric Pressures"; Ahmed, "Affective Economies."

9. Foucault, *Security, Territory, Population*, 48–49.

10. In my usage, "queerness" is neither an identity category nor a locus of resistance. I do not define queerness in relation to forms of sovereignty that would render either of these meanings legible. The queer is simply the unformed potential of vital transition and differentiation, which energizes the plasticity of power. I pose

life's queer potential against "pro-life" moralisms (whether applied to abortion, euthanasia, biotechnologies, or animal ethics) that reduce life to the able-bodied form of the reproductive, vertebrate body. For more on life's queer potential, see Colebrook, "How Queer Can You Go?," 22–23. However, I do not attempt to use this view of life to queer theory itself, as Colebrook provocatively does.

11. Cohen, "The Politics of Viral Containment," 15–16.

12. Chen, *Animacies*, 3.

13. Hayward, "Lessons from a Starfish"; Hird, "Animal Trans"; Halberstam, *The Queer Art of Failure*, 27–52.

14. Puar, "Prognosis Time," esp. 166.

15. Wald, *Contagious*, 2, 37. Wald's work builds on earlier studies including Rosenberg, *Explaining Epidemics*; Briggs with Mantini-Briggs, *Stories in the Time of Cholera*.

16. Today vaccinia virus used in smallpox vaccine is grown in lab cultures rather than on cows. The cultured sera are still derived from earlier bovine stocks and are tested on mice, rabbits, rhesus macaques, and humans. See Weltzin et al., "Clonal Vaccinia Virus."

17. Stepan, *Eradication*, 8.

18. Jerne, "Towards a Network Theory of the Immune System"; Jerne, "Nobel Lecture."

19. Anderson, "Natural Histories of Infectious Disease," 58–59.

20. Dubos, *So Human an Animal*, 24–25.

21. At a press conference in the lead-up to the Iraq War, Rumsfeld famously distinguished three types of threat, implying that Iraq fit into the third and most inscrutable category: (1) "known knowns . . . there are things we know we know"; (2) "known unknowns . . . things we do not know"; and (3) "unknown unknowns . . . the ones we don't know we don't know. . . . It is the latter category that tends to be the difficult one." Department of Defense News Briefing, February 12, 2002, http://www.defense.gov/transcripts/transcript.aspx?transcriptid=2636.

22. Lakoff, "The Generic Biothreat."

23. For key biosemiotic theories, see Uexküll, *A Foray into the Worlds of Animals and Humans*; Kohn, *How Forests Think*; Rueckert, "Literature and Ecology."

24. Leys, "The Turn to Affect."

25. Mawani, "Atmospheric Pressures"; Silva, *Toward a Global Idea of Race*.

26. On affective labor as one type of immaterial labor, see Hardt and Negri, *Empire*, 293; on affect as surface intensity, see Jameson, "Postmodernism," 64. For a critique of theories that fail to specify the content of affect, see Khanna, "Touching, Unbelonging, and the Absence of Affect."

27. Thrailkill, *Affecting Fictions*, 15–16.

28. For an outline of five major approaches to affect, see Thrift, *Nonrepresentational Theory*, 223–25. On the relation of affect and emotion, see Puar and Pellegrini, "Affect"; Puar, "Prognosis Time," 161–62. For key vitalist and species-critical approaches to affect, see Massumi, *Parables for the Virtual*; Latour, "How to

Talk about the Body?"; Parreñas, "Producing Affect"; Hustak and Myers, "Involutionary Momentum."

29. Massumi, *Parables for the Virtual*, 1–23.

30. On alimentation and the affective politics of empire, see Roy, *Alimentary Tracts*.

31. Kraut, *Silent Travelers*, 3. On attributions of disease risk to racialized groups in the United States, see further Markel, *Quarantine!*; Molina, "Medicalizing the Mexican"; Ong, "Making the Biopolitical Subject"; Shah, *Contagious Divides*; Stern, *Eugenic Nation*; Wailoo, *Dying in the City of the Blues*; Wald, *Contagious*. On tropical medicine and public health colonialism in the US and British empires, see Anderson, *Colonial Pathologies*; Anderson, *The Cultivation of Whiteness*; Gussow, *Leprosy, Racism, and Public Health*; Arnold, *Colonizing the Body*; Arnold, *Imperial Medicine and Indigenous Societies*; Ballhatchet, *Race, Sex, and Class under the Raj*; Briggs, *Reproducing Empire*; Lindsay-Poland, *Emperors in the Jungle*; Pati and Harrison, *Health, Medicine, and Empire*; Stern, "The Public Health Service in the Panama Canal"; Vaughan, *Curing Their Ills*.

32. Abernathy, "Bound to Succeed"; Stepan, *Eradication*.

33. Gussow, *Leprosy, Racism, and Public Health*, 131.

34. Patient activism at the Carville National Leprosarium, Louisiana, was instrumental in convincing medical authorities to replace the term "leprosy" with "Hansen's disease" in order to combat the stigma associated with the former term. On this and other aspects of the "destigmatization theory" elaborated by patient activists, see Gussow, *Leprosy, Racism, and Public Health*, 12–15. "Leprosy" is still the common medical designation for the disease outside the United States. "Hansen's disease" (named for Norwegian physician Armauer Hansen, who theorized the bacteriology of *Mycobacterium leprae*) poses its own difficulties as it idealizes scientific medicine within a colonial logic of discovery, in the process subjugating other extant knowledges of the disease in distinct medical traditions. However, I have chosen to use the imperfect term "Hansen's disease" to support the ongoing campaigns that critique the stigmatization of patients. My occasional references to the pejorative noun "leper" and adjective "leprous" refer only to stereotyped depictions of the disease.

35. By this time decolonial nationalist movements and the rise of working classes in liberal democracies undermined the financial and political logics of mass territorial expansionism. The wartime era instead conjoined privileges of imperial alliance (the Monroe Doctrine and Open Door policies, for example) and military-industrial stimulus that channeled expanding US financial capital into wartime manufacturing. I draw here on the Marxist theories of US empire driven by finance capital in Hudson, *Super Imperialism*; Arrighi, *The Long Twentieth Century*; Desai, *Geopolitical Economy*.

36. For a brilliant reading of the rendering of animal bodies in media and capital circuits, see Shukin, *Animal Capital*.

37. Committee on the Establishment of a Cardiovascular Primate Colony, Na-

tional Advisory Heart Council Committee, Meeting Minutes, September 25, 1957, 19, Box 43, Folder "National Advisory Heart Council—Primate Colony, 1957–8," George E. Burch Papers, National Library of Medicine, Bethesda, Maryland.

38. Stepan, *Eradication*, 8; Cooper, "Preempting Emergence," 114–15.

39. King, "Security, Disease, Commerce," esp. 775, 777; Cooper, "Preempting Emergence," 113.

40. Arvin posits a politics of regeneration, rather than binary decolonization, in native Hawaiian responses to the imperial incorporation of the "Polynesian" by colonial racial sciences. See Arvin, "Pacifically Possessed."

41. Fanon, "Medicine and Colonialism," 126–27. Fanon is informed by Freud's account of anxiety as privileged affective relation of mind and body. Fanon's Cartesian use of the term is thus not continuous with vitalist strains of affect theorizing that take affect as unformed bodily intensity. On the Freudian theorization of affect, see Khanna, "Touching, Unbelonging, and the Absence of Affect," 218–20.

42. Fanon, *The Wretched of the Earth*, 6, 182.

43. Fanon, *The Wretched of the Earth*, 19.

44. Gandhi, *The Story of My Experiments with Truth*. See also the brilliant analysis of Gandhi's poetics of fasting in Roy, *Alimentary Tracts*, chapter 3.

45. Gandhi, *Hind Swaraj*, 95, 53.

46. Foucault, *Security, Territory, Population*, 200–201.

Chapter 1. "An Atmosphere of Leprosy"

1. Mouritz, *The Path of the Destroyer*, 9 (prologue), 24.

2. Roy, *Alimentary Tracts*, 7.

3. On US empire, sovereignty, and naturalization of white business interests abroad, see Lewis, "Naturalizing Empire."

4. While the US figure of the Haitian *zombi* is generally dated later, to the publication of William Seabrook's 1929 novel *The Magic Island*, the production of the Hawaiian Hansen's patient as embodying a liminal zone of living death and species hybridity anticipates Seabrook's racial framing of the zombie as spiritually tainted and contagious. There was a close connection between US imperial framings of tropical islands, racialized populations, and disabled embodiments that anticipated the horror of the zombi in the figure of the leper. See Seabrook, *The Magic Island*. On the rendering of the zombi as colonial trope, see further Dayan, *Haiti, History, and the Gods*, 36–37.

5. Frazier, introduction, ix.

6. Moblo, "Blessed Damien of Moloka'i"; Silva and Fernandez, "Mai Ka 'Āina O Ka 'Eha'eha Mai," 76.

7. Gavan Daws, cited in Edmund, *Leprosy and Empire*, 147; Silva and Fernandez, "Mai Ka 'Āina O Ka 'Eha'eha Mai," 85.

8. Silva and Fernandez, "Mai Ka 'Āina O Ka 'Eha'eha Mai," 82.

9. *In re Kaipu*, December 21, 1904, in *Report of Causes Determined in the United States District Court for the District of Hawai'i*, 215–29; "Great Test Case Ended by

Death," *Maui News*, April 20, 1907, http://distantcousin.com/obits/hi/1900/1907/Apr/MauiNews/20/Kaipu.html.

10. Anderson, *Colonial Pathologies*, 158–79; Gussow, *Leprosy, Racism, and Public Health*, 131; Lara, "Leprosy Research in the Philippines."

11. J. F. Curtis, Office of the Secretary of the Treasury Department of the United States, "Transportation of Lepers in Interstate Traffic: An Amendment to Interstate Quarantine Regulations," May 15, 1912.

12. Silva, *Aloha Betrayed*, 24–27.

13. Moblo, *Defamation by Disease*. Among popular histories of Moloka'i, John Tayman's *The Colony* attempts to tell histories of patient activism at Moloka'i. However, Kalaupapa residents including Olivia Breitha have claimed that Tayman's title, cover illustration, use of the term "leprosy," erasure of patient activism, and looseness with historical events and citations sensationalized the story. I agree that Tayman's work does have a number of weaknesses, including these and its idealized depiction of Father Damien. Although an earlier version of one section of this chapter included some citations from Tayman, I have not quoted him here out of deference to the former settlement residents and in order to more accurately present the historical narrative. See Tayman, *The Colony*, 319–20; Moblo, "Review," 889–90; Michael Wilson, "Book on Leprosy Settlement Draws Fire," *New York Times*, March 27, 2006, http://www.nytimes.com/2006/03/27/books/27wils.html.

14. Stoddard, *The Lepers of Molokai*, 18.

15. Shah, *Contagious Divides*, 99–100; Gussow, *Leprosy, Racism, and Public Health*, 111–29.

16. Palmer, *The Human Side of Hawaii*, xi.

17. "Shall We Annex Leprosy?," *The Cosmopolitan* 24 (1897); Morrow, "Leprosy and Hawaiian Annexation."

18. *A Report Relating to the Origin and Prevalence of Leprosy in the United States*, 119. Further references in text.

19. References to the Caribbean highlight a parallel British discourse on Hansen's disease transmission from Asia to the Americas, which colonial officials traced along the routes of the indentured labor system connecting India to the West Indies.

20. Shah, *Contagious Divides*, 100.

21. Gussow, *Leprosy, Racism, and Public Health*, 123–24.

22. On the history of the magic lantern, see Castle, *The Female Thermometer*.

23. Gilman, *Disease and Representation*, 1–2.

24. On the medical animalization of Hansen's disease, see further Edmund, *Leprosy and Empire*, 38.

25. Garland Thomson, "Seeing the Disabled," 346–47.

26. Appadurai, "The Colonial Backdrop," 5.

27. On pre-germ theory understandings of disease in the United States, see Tomes, *The Gospel of Germs*, 2–5.

28. Hitt, "Leprosy," 331.

29. See Prashad, *The Karma of Brown Folk*, 21–45.

30. Garland Thomson, "Seeing the Disabled," 336, 338, and 342.

31. Subcommittee on Pacific Islands and Porto Rico, *Hawaiian Investigation Report*, 73–75.

32. Ryan, *Picturing Empire*, 17, 19.

33. Edmund, *Leprosy and Empire*, chap. 3.

34. Mouritz, *The Path of the Destroyer*, 9 (prologue).

35. Gibson, "The Lepers and Their Home on Moloka'i," *Nuhou*, March 14, 1873, 188–89, quoted in Greene, *Exile in Paradise*, 16–17.

36. Arvin, "Pacifically Possessed," 20.

37. Streeby, *American Sensations*. Sensationalism pervaded many types of writing, including the realist and naturalist movements influential in constructing a sense of the deterministic relationship between environment and individual in the early twentieth century. Accounts of the "horrors of Moloka'i"—written by canonized authors including Charles Warren Stoddard and Robert Louis Stevenson—often contrasted Hansen's patients with virtuous Catholic missionaries. Such literary strategies allowed authors to, on the one hand, accept the germ theory of the spread of the disease and, on the other, propose that disease was a spiritual taint or burden—one that needed the remedy of Progressive reform. Such a portrayal is apparent in Stevenson, *Father Damien*.

38. See Shakespeare, "The Social Model of Disability."

39. Stoddard, *The Lepers of Molokai*, 58–59. Further references in text.

40. Shah, *Contagious Divides*, 99–100.

41. Lye, *America's Asia*, 36, 94, 274n98.

42. Arvin, "Pacifically Possessed," 118.

43. On the prevalence of Hansen's disease in India, see Kakar, "Medical Developments and Patient Unrest," 191–92, 200, 204.

44. Halberstam, *Skin Shows*, 2.

45. Tomso, "The Queer History of Leprosy and Same-Sex Love," 761.

46. Moblo, "Institutionalizing the Leper," 234.

47. London, "Koolau the Leper," 18–19. Further references in text.

48. In particular, London fears "male degeneracy" and sides with idealizers of an essentialized masculinity of physical power. See Hoganson, *Fighting for American Manhood*, 133–55; Bederman, *Manliness and Civilization*.

49. Gussow, *Leprosy, Racism, and Public Health*, 123–24.

50. Mitchell and Snyder, "Narrative Prosthesis and the Materiality of Metaphor."

51. On London's masculinism, see Forrey, "Male and Female in London's *The Sea Wolf*." Indigenous peoples were often feminized by depicting them as receptive to disease. See, for example, Mouritz, *The Path of the Destroyer*, 9 (prologue): "Civilization blights aboriginal races, the Hawaiian is no exception; the harsh side of civilization kills by kindness, it may resemble the Greeks bearing gifts, in contact with primitive races it is treacherous."

52. Silva, *Toward a Global Idea of Race*, xix.

53. McClintock (*White Logic*, 137) and Labor ("Jack London's Pacific World," 211) also see "Koolau" as derivative of London's other heroes.

54. Memmi argues that dependency is necessarily reciprocal at the scales of the body and the nation, and argues that neocolonial economic relationships are relations of "reciprocal need." Here dependence is seen as "the expression of the permanent reciprocity that, because of their needs, exists between most of the members of a group: dependence of the weak on the strong, but also of the strong on the weak." As such, one might argue that colonial violence is the inaugurating event that enforces the colonizer's need to be needed. However, Memmi's purely psychoanalytic model fails to engage with either the unequal ability of one party to exert power over another or the spatial materiality of dependencies. See Memmi, *Dependence*, 33, 65–66, and 154.

55. Lye, *America's Asia*, 8.

56. Kaluaikoʻolau, *The True Story of Kaluaikoʻolau as Told by His Wife, Piʻilani*, 17.

57. Kaluaikoʻolau, *The True Story of Kaluaikoʻolau as Told by His Wife, Piʻilani*, 33.

58. Kaluaikoʻolau, *The True Story of Kaluaikoʻolau as Told by His Wife, Piʻilani*, 47.

59. Breitha, *Olivia*, 13.

60. Breitha, *Olivia*, 98.

61. Vaughan, *Curing Their Ills*, 77–99; Moran, *Colonizing Leprosy*; Anderson, *Colonial Pathologies*, chap. 6; Kakar, "Medical Developments and Patient Unrest."

62. Moran, *Colonizing Leprosy*, 194.

63. Silva, *Aloha Betrayed*, 148–50; Silva and Fernandez, "Mai Ka ʻĀina O Ka ʻEhaʻeha Mai," 93–94.

64. Agamben, *Homo Sacer*, 154–59.

65. Edmund, *Leprosy and Empire*, 92.

66. Turse, "Experimental Dreams, Ethical Nightmares," 154–55.

67. Turse, "Experimental Dreams, Ethical Nightmares," 147.

68. Kakar, "Leprosy in British India," 215–16.

69. Parascandola, "Chaulmoogra Oil and the Treatment of Leprosy."

70. Kakar, "Medical Developments and Patient Unrest," 191–92, 200, 204.

71. Turse, "Experimental Dreams, Ethical Nightmares," 149.

72. Foucault, *Discipline and Punish*, 199.

73. Agamben, *Homo Sacer*, 168–69.

74. Mbembe, "Necropolitics," 21.

75. Ahuja, "Abu Zubaydah and the Caterpillar," 143.

76. Puar, "Prognosis Time," 163.

77. Medevoi, "Global Society Must Be Defended."

78. Burnett, "The Edges of Empire and the Limits of Sovereignty," 800.

Chapter 2. Medicalized States of War

1. Reverby, *Examining Tuskegee*, 139.

2. As in the US-occupied Philippines, public health intervention was undertaken under military occupation authority. This situation differed from US-sponsored public health efforts in Costa Rica and El Salvador, which were largely funded by philanthropists such as Rockefeller. Nonetheless, there was a widespread circulation

of expert knowledge between US and Central American government health officials, military health, and the philanthropists. See Palmer, "Central American Encounters with Rockefeller Public Health," 326; Chamberlain, *Twenty-Five Years of American Medical Activity.*

3. Salvatore, "Imperial Mechanics," 663; Chase Smith, "Bawdy Amusements and the Undergarments of 'Progress' at San Diego's 1915–1916 Panama-California Exposition," unpublished manuscript, 2008; Stern, "The Public Health Service in the Panama Canal," 675–80.

4. Abernathy, "Bound to Succeed," 3–4.

5. Biesanz and Biesanz (*The People of Panama*, 67–92, 221–31) visited the isthmus in the 1940s. They claim that American insularity, an oppositional settler relationship to the US government, and institutional and economic structures helped maintain in the Canal Zone a form of racial apartheid stricter than in the Jim Crow South. This chapter suggests that fears of contagion were also major contributors to the forces tending toward racial and gender apartheid at the Zone.

6. Frenkel, "Geographical Representations of the 'Other,'" 89.

7. Lindsay-Poland, *Emperors in the Jungle*, 33.

8. Stern, "The Public Health Service in the Panama Canal," 675–76.

9. Lindsay-Poland, *Emperors in the Jungle*, 30.

10. Lindsay-Poland, *Emperors in the Jungle*, 11.

11. See Lindsay-Poland, *Emperors in the Jungle*, 16; LeFeber, *The Panama Canal*; Major, *Prize Possession.*

12. Abernathy, "Bound to Succeed," 156.

13. In 1946, women's political parties succeeded, after decades of struggle, in securing constitutional recognition of suffrage in Panamá. Having been offered limited suffrage by the right-wing nationalist government of Arnulfo Arias Madrid in the 1941 constitution, the major feminist party, the Partido Nacional Feminista, rejected the nationalist platform, joining with the Progressive Front that would succeed in expanding the citizenship rights of women and racial minorities in 1946. Yet in the preceding years, when the cooperative venereal disease program was initially instituted, one of the key complaints of the party against the nationalist platform was "the authoritarian character of the government" under Arias that sought "control of the domestic population." This included its notorious eugenic racial targeting in deportation and sterilization initiatives. See Serra, "El Movimiento Sufragista en Panamá y la Construcción de la Mujer Moderna," 123; Herrera, *La Discriminacion de la Mujer en Panamá*, 57–58. Unless otherwise noted, all English translations of texts originally written in Spanish in this chapter are by the author.

14. Brandt, *No Magic Bullet*, 72.

15. On outlawing prostitution, see Brandt, *No Magic Bullet.*

16. Briggs, *Reproducing Empire*, 21.

17. Panama Canal Department, United States Army, *Control of Venereal Disease and Prostitution*, 9–10. Further references in text.

18. Brandt, *No Magic Bullet*, 165.

19. Brandt, *No Magic Bullet.*

20. Ferguson, "Our Normative Strivings."

21. Briggs, *Reproducing Empire*; Rosen, *The Lost Sisterhood*.

22. Panama Canal Zone Health Department, "Report," 260.

23. Siu, *Memories of a Future Home*, 115–16.

24. Cabezas, *Economies of Desire*, 20, 22.

25. On the treatment of syphilis through chemical prophylaxis, see Worms, "Prophylaxis of Syphilis by Locally Applied Chemicals."

26. Shah, "Prostitution, Sex Work, and Violence," 795–96.

27. See Hua, *Trafficking Women's Human Rights*.

28. "Propose Ban on Brothels in Main Cities," *Panama Tribune*, November 30, 1947, 1.

29. "RP Concerned over Army Ban in Main Cities: President Says Government Plans Drastic Measures to Curb Vice," *Panama Star and Herald*, May 1, 1946, 1. See also "Se Discutió el Problema de la Prostitución: Sobre Ésto Hablaron el Ministro Vallarino y el General Homer," *La Estrella de Panamá*, May 3, 1946, 1; "Zone Co-operation with Panama on RP Vice Problem," *Panama American*, May 14, 1946, 1.

30. Brandt, *No Magic Bullet*, 177.

31. "Las recogidas nocturnas efectúanse en Colón causan pánico en el comercio e inquietud entre las personas de bien," *El Panamá América*, June 28, 1947, 1, 4. Many thanks to Zulema Diaz for assisting with the translation of this article.

32. "Panama City's Bars Plan Week's Strike against Army's Law," *Panama American*, May 9, 1946, 1.

33. "El Porcentaje de Contagion Venereo Ha Bajado Bastante," *El Panamá América*, February 22, 1947, 1, 11.

34. Guillermo García de Paredes, "El Control de las Enfermedades Venéreas en Panamá," La XII Conferencia Sanitaria Panamericana, Caracas, Venezuela, 1947. Pamphlet Vol 5937/25972210R, History of Medicine Division, National Library of Medicine, Bethesda, Maryland, 7.

35. "Fueron Recogidas en Colón Numerosas Mujeres, de las Cuales 12 Enfermas," *El Panamá América*, May 23, 1947, 1.

36. "Brothel" appears in Fanon, *The Wretched of the Earth*, trans. Farrington, 154, while "bordello" appears in Fanon, *The Wretched of the Earth*, trans. Philcox, 102.

37. Condé, *Tree of Life*, 11.

38. McPherson, "Rioting for Dignity."

39. Frederick (*Colón Man a Come*, 198) calls for "a mythographic approach" to the study of women laborers who worked in the Canal Zone under US authority at the turn of the twentieth century. According to Frederick, women's experiences "are not adequately represented in *any* isthmian narrative."

40. Sinán was one of Panamá's main poetic vanguards. He adapted the Central American traditions of *antiimperialismo* and *criollismo* to a surrealist global vision of the violences of empire and war. See Unruh, *Latin American Vanguards*, 2–3; Boland Osegueda, "The Central American Novel," 165.

41. Franco, *The Modern Culture of Latin America*, 270–71. Franco emphasizes Sinán's interest in "his native land" despite acknowledging that "his use of fan-

tasy and avant-garde techniques and . . . cosmopolitan outlook" separate him from other nationalist writers.

42. Sinán, *Plenilunio*, 19. Further references in text.

43. The original Spanish text reads, "Las arcas del Tío Sam se derramaban para dines de guerra. . . . Nos caían en el Istmo algunas gotas—muchedumbre de gotas que muchos recogían avaramente, sedientos. . . . Yo me cegué. No tuve escrúpulos en gastar mi dinero con prostitutes. . . . ¡Habían llegado tantas al Istmo! ¡Mexicanas, cubanas, argentinas . . . de todas partes las había!"

44. In Spanish: "Todo caía encharcado: La Moral, en pollera, cantaba el Himno dentro de los burdeles. . . . Aquello era la exaltación del egoísmo, del sexo, de la prostitución."

45. Aguilera-Malta, *Canal Zone*, 13. Further references in text.

46. In Spanish: "Las cantinas y los cabarets abrían sus ojos imantados. Sacaban afuera sus ferias de carne. Brindaban su canción de botellas. Se emborrachan, frente a las cementerios. Vomitaban piltrafas humanas, cuando el sol barría la madrugada."

47. Walrond's biography is one of multiple migrations, as he moved from British Guyana to Barbados to Panamá before joining Marcus Garvey in New York during the Harlem Renaissance and then producing his final works in England and Spain. In Panamá, Walrond worked for the Canal Commission's Health Department as a clerk. His stories dealing with subaltern health problems based on drought and malnutrition were published during his time in Harlem, where he became an editor of *Negro World* and published his most famous stories of Panamá collected in the volume *Tropic Death*. Because he wrote in English, Walrond is often figured outside of the literary canon of Panamá despite his lengthy residence in the country.

48. Walrond, "Godless City," 163. Further references in text.

49. Walrond, "Morning in Colón," 312.

50. Noting the complicated relationships of Caribbean writers to modernist form given its immersion in colonial projects, Simon Gikandi (*Writing in Limbo*, 5) argues, "Caribbean modernism has evolved out of an anxiety toward the colonizing structure in general and its history, language and ideology in particular." Gikandi further locates the Caribbean traditions of modernism and creolization as a "counter-discourse away from outmoded and conventional modes of representation associated with colonial domination and colonizing cultural structures."

Chapter 3. Domesticating Immunity

1. "First American Monkey Colony Starts on Puerto Rico Islet," 26. For a key example of the racialized US discourse on Hindu patriarchy, see Mayo, *Mother India*. For a critique of the deployment of this discourse by the British in India, see Spivak, "Can the Subaltern Speak?," esp. 92.

2. Mieth describes dreams that haunted her in the years following the trip to Puerto Rico. In these dreams, the frustrated monkey queries the artist about the

possibility of leaving its stranded island world; Mieth responds with an invocation to wait "until things get better." In this dream of interspecies communication, the photographer and research monkey share a sense of alienation bred by modern human conflict. See Flamiano, "Meaning, Memory, and Misogyny," 26.

3. Glick, "Ocular Anthropomorphisms," 106–7.

4. Burt, *Animals in Film*, 104.

5. Clause, "The Wistar Rat as a Right Choice"; Kohler, *Lords of the Fly*; Clarke, *Disciplining Reproduction*.

6. Martin, *Flexible Bodies*, 31.

7. Macfarlane and Worboys, "The Changing Management of Acute Bronchitis in Britain," 47. See also Weatherall, *In Search of a Cure*.

8. Serlin, *Replaceable You*, 3.

9. Martin, *Flexible Bodies*, 23–31.

10. Coughlan, "Science Moves In on Viruses," 122.

11. Wilson, "A Crippling Fear," 464–65, 468, 495n13.

12. Wilson, "A Crippling Fear," 466.

13. Eggers, "Milestones in Early Poliomyelitis Research." See further Landsteiner and Popper, "Uebertragung der Poliomyelitis Acuta auf Affen."

14. Paul, *A History of Poliomyelitis*, 243; Rogers, *Dirt and Disease*, 23–25. As Rogers puts it, "drawing analogies between a disease in humans and symptoms in animals is often risky," and in the case of polio, may have set back research for decades until researchers in the 1940s and 1950s began to focus on the oral-intestinal pathway.

15. Flexner and Amoss, "The Relation of the Meninges and Choroid Plexus to Poliomyelitic Infection," 79.

16. Oshinsky, *Polio*, 56–58.

17. This section offers an abbreviated discussion of the history of US primate importation and breeding linking the United States, Puerto Rico, India, and central Africa. Most of the information concerning India and Africa in this section comes from the George E. Burch Papers at the National Library of Medicine in Bethesda, Maryland, henceforth cited as George E. Burch Papers. For more in-depth discussion of this transnational history and detailed citation of my archival sources, see Ahuja, "Macaques and Biomedicine"; Ahuja, "Notes on Medicine, Culture, and the History of Imported Monkeys in Puerto Rico."

18. Rawlins and Kessler, "The History of the Cayo Santiago Colony," 24.

19. "First American Monkey Colony Starts on Puerto Rico Islet," 26.

20. "First American Monkey Colony Starts on Puerto Rico Islet," 26.

21. Carpenter, "Rhesus Monkeys (*Macaca Mulatta*) for American Laboratories," 285.

22. Stepan, *Picturing Tropical Nature*.

23. Locke, "Peopling an Island with Gibbon Monkeys," 290.

24. Rawlins and Kessler, "The History of the Cayo Santiago Colony," 22.

25. Haraway, *Primate Visions*, 260; Prashad, *The Darker Nations*, 11, 43.

26. G. E. Burch, "A Proposal for a National Cardiovascular Primate Station," 1, February 3, 1958, Box 43, Folder "National Advisory Heart Council—Primate Colony, 1957–8," George E. Burch Papers.

27. Victoria Harden, "Interview with Dr. William I. Gay," Office of NIH History, July 15, 1992, http://history.nih.gov/NIHInOwnWords/docs/gay_01.html; Committee on the Establishment of a Cardiovascular Primate Colony, National Advisory Heart Council, Meeting Minutes, 2, 20–21, September 25, 1957, Box 43, Folder "National Advisory Heart Council—Primate Colony, 1957–8," George E. Burch Papers.

28. Haraway, *Primate Visions*, 121. These institutions were renamed the National Primate Research Centers following the 9/11 attacks. The eight centers—located in Beaverton, Oregon; Atlanta, Georgia; San Antonio, Texas; Southborough, Massachusetts; Davis, California; Covington, Louisiana; Madison, Wisconsin; and Seattle, Washington—have housed forty-five species of primates. Cayo Santiago was incorporated into the Caribbean Regional Primate Center administered through the University of Puerto Rico.

29. In later years, more efficient cell culture production techniques cut that number nearly in half. See Haraway, *Primate Visions*, 413n34.

30. Paul Weiss, quoted in Haraway, *Primate Visions*, 121–22.

31. Committee on the Establishment of a Cardiovascular Primate Colony, Meeting Minutes, 17.

32. Committee on the Establishment of a Cardiovascular Primate Colony, Meeting Minutes, 18.

33. Committee on the Establishment of a Cardiovascular Primate Colony, Meeting Minutes, 19.

34. Committee on the Establishment of a Cardiovascular Primate Colony, Meeting Minutes, 25–26.

35. Haraway, *Primate Visions*, 236–43.

36. Harlow, "The Nature of Love."

37. Harlow, "The Evolution of Harlow Research"; Haraway, *Primate Visions*, 231.

38. Longmore, "'Heaven's Special Child.'"

39. Lundblad, *Birth of a Jungle*, chap. 5; Glick, "Ocular Anthropomorphisms."

40. See further Kim, "The Racial Triangulation of Asian Americans."

41. Shell, *Polio and Its Aftermath*, 129–38.

42. Mitchell and Snyder, "Narrative Prosthesis and the Materiality of Metaphor."

43. Theodore Ruch, quoted in Committee on the Establishment of a Cardiovascular Primate Colony, Meeting Minutes, 27.

44. See my longer discussion of this history in Ahuja, "Notes on Medicine, Culture, and the History of Imported Monkeys in Puerto Rico."

45. Phoebus, Roman, and Herbert, "The FDA Rhesus Breeding Colony at La Parguera," 157; Gonzáles-Martínez, "The Introduced Free-Ranging Rhesus and Patas Monkey Populations."

46. G. N. Ramírez, "Urge Acción Oficial ante la Amenaza de los Monos," *El Nuevo Día*, May 27, 1998; A. N. Alfaro, "Cultivando para los Monos," *El Nuevo Día*, July 27, 2008.

47. Anti-paranormal theorist Benjamin Radford has argued that the myth of el chupacabras can be entirely explained by the influence of the film *Species*, released around the time of the first sightings. This argument has been received skeptically within online paranormal communities and cannot in itself explain the wide dissemination of the cryptid within and outside Puerto Rico. See Radford, *Tracking the Chupacabra*; Lauren Coleman, "Chupacabras Solution: Neither New Species, Nor Silly," *Cryptomundo*, May 21, 2010, http://www.cryptomundo.com/cryptozoo-news /radford-theory/.

48. Corrales, *Chupacabras and Other Mysteries*, 115–16.

49. Derby, "Imperial Secrets," 310.

50. Derby, "Imperial Secrets," 294.

51. McCaffrey, *Military Power and Popular Protest*, 138–46.

52. Jordan, *El Chupacabra*, 4.

53. "Advierten Peligro de Monos Rhesus," wapa.tv, December 19, 2008, http:// www.wapa.tv/noticias.php?nid=20081219214708; "Nuevo Secretario del DRNA Revisa Plan de Captura de Monos," Associated Press, January 21, 2009.

54. Burch, "A Proposal for a National Cardiovascular Primate Station," 1.

55. On the neoliberal legacies of scientific discourses of primate likeness, see Hua and Ahuja, "Chimpanzee Sanctuary."

56. Annan, foreword.

57. Lowe, "The Intimacies of Four Continents."

58. Potter, *Green Is the New Red*.

Chapter 4. Staging Smallpox

1. Preston, foreword, 12.

2. These Left critiques have nonetheless been formative for my understanding of continuities in the history and practice of US imperial interventionism. See especially Johnson, *Blowback*; Chomsky, *Rogue States*. Melani McAlester extends this work with her complex reading of the interpenetration of missionary, multicultural, and military worldviews in the propagation of the war. See McAlester, *Epic Encounters*, 266–309; McAlester, "Rethinking the 'Clash of Civilizations.'"

3. Appadurai, *Fear of Small Numbers*, 81.

4. The fact that the US military circulated stories of Saddam's Viagra use, and that tabloids and the television program *South Park* joked that he was a homosexual, only confirm the anxiety around his military force. In response, Saddam becomes the object of the predictable, heteronormative strategy of emasculation.

5. Lakoff, "The Generic Biothreat."

6. Cooper, "Preempting Emergence," 115, emphasis in original.

7. King, "Security, Disease, Commerce," 763.

8. George W. Bush, quoted in Bartlett et al., "Smallpox Vaccination in 2003," 883.

9. Commission on the Prevention of WMD Proliferation and Terrorism, *World at Risk*, 24.

10. King, "Security, Disease, Commerce."

11. Five of twenty-two victims died in the 2001 anthrax attacks, which were carried out through the US mail. The FBI suspects that the late Bruce Ivins, a former microbiologist with state security clearance and access to anthrax stocks, carried out the attacks. However, state and media immediately suspected a foreign source for the attacks. George Tenet's memoirs allege that al-Qaeda hired two separate bioweapons engineers with knowledge of anthrax to build a bioweapons program in Afghanistan. See Tenet, *At the Center of the Storm*, 278.

12. Kraut, *Silent Travelers*, 21–25, 82; Shah, *Contagious Divides*, 57–63.

13. Crosby, *The Columbian Exchange*, 42–63. Crosby's initial figures were disputed, and he later retracted his claims that 30 percent of indigenous peoples were killed by smallpox, making European conquest inevitable. Some medical-historical accounts suggest that a Spanish swine flu may have been the key settler disease rather than smallpox. See Kraut, *Silent Travelers*, 21–25. Yet the exaggerations of smallpox's world-historical role in culling human populations continues to lend urgency to the neoconservative figuration of smallpox as a reemerging disease.

14. Thornton, *American Indian Holocaust and Survival*, 78–81; Peter d'Errico, "Jeffrey Amherst and Smallpox Blankets," 2010, http://www.nativeweb.org/pages /legal/amherst/lord_jeff.html.

15. Tucker, *Scourge*, 12.

16. McNeill, "Patterns of Disease Emergence in History," 30.

17. Henderson et al., "Smallpox as a Biological Weapon," 2128. See also Segelid, "Smallpox Revisited," W5–W11.

18. Henderson, *Smallpox*, 43.

19. Byrd, *The Transit of Empire*.

20. Preston, *The Demon in the Freezer*; Alibek with Handleman, *Biohazard*.

21. David Willman, "Selling the Threat of Bioterrorism," *Los Angeles Times*, July 1, 2007, http://articles.latimes.com/2007/jul/01/nation/na-alibek1.

22. Miller, Engelberg, and Broad, *Germs*, 294–99.

23. Preston, *The Demon in the Freezer*, 165–91.

24. Willman, "Selling the Threat of Bioterrorism"; Lederberg, "Appendix C," 236.

25. The best evidence of Iraq's unconventional weapons would have been the long record of international exports, approved by the United States, of military and dual-use items. The CDC and the American Type Culture Collection had directly exported bioagents to its ally Iraq during the Iran-Iraq War (the first Gulf War). Some signatories of the neoconservative Project for a New American Century's Iraq letters of 1998, including Donald Rumsfeld, had carried out negotiations with Saddam and helped build Iraq's weapons programs. Despite quiet official statements against Iraq's use of chemical weapons, the United States supported Saddam before and after Iraq's gas attacks of 1983–88 on Iran and the Kurds, going as far as to block a UN resolution condemning Iraq's use of chemical agents. The US Commerce Department approved an estimated seven hundred transfers of dual-use materials (including anthrax) to Iraq during the Iran conflict, and various European and Asian chemical and petroleum companies legally sold Iraq chemical agents. The United

States only ended its support of Saddam in 1990 upon Iraq's invasion of Kuwait. See Pythian, *Arming Iraq*; Battle, *Shaking Hands with Saddam Hussein*.

26. This logic was evident as early as 1963, when Norman Podhoretz's racist essay "My Negro Problem—And Ours" (see Gerson, *The Essential Neoconservative Reader*) posited a racialized uncertainty to social relations that could be addressed only through division and securitization rather than social democracy.

27. See Gerson, *The Essential Neoconservative Reader*, especially the essays by Norman Podhoretz, Daniel Patrick Moynihan, Ruth Wisse, and Irving Kristol.

28. Project for the New American Century, "Statement of Principles," June 3, 1997, https://web.archive.org/web/20050205041635/http://www.newamericancentury.org/statementofprinciples.htm.

29. Donnelly, *Rebuilding America's Defenses*, iv.

30. Schelling, *The Diplomacy of Violence*; Donnelly, *Rebuilding America's Defenses*, 3.

31. US Strategic Command, "Post–Cold War Deterrence," quoted in Chomsky, *Rogue States*, 6.

32. Derrida, *The Beast and the Sovereign*, 89. Derrida draws on the critique of the Clinton interventions in Iraq in Chomsky, *Rogue States*, esp. 7 and 13.

33. Podhoretz, *World War IV*.

34. Suskind, "Faith, Certainty, and the Presidency of George Bush." See also Danner, "Words in a Time of War," 17.

35. Project for the New American Century, "Letter to President Clinton on Iraq," January 26, 1998, https://web.archive.org/web/20131021171040/http://www.newamericancentury.org/iraqclintonletter.htm, emphasis added.

36. Project for the New American Century, "Letter to Gingrich and Lott on Iraq," May 29, 1998, https://web.archive.org/web/20120313105822/http://newamericancentury.org/iraqletter1998.htm.

37. Chomsky, *Rogue States*, 13–14.

38. McAlester, *Epic Encounters*, 258.

39. Galison, "Removing Knowledge," 231.

40. Masco, "'Sensitive but Unclassified.'" On the open secret and the management of public fear, see further Ahuja, "Abu Zubaydah and the Caterpillar," 136–37.

41. Masco, "Lie Detectors."

42. Miller's obituary tribute reports Patrick "launched a one-man campaign to warn government officials, policy experts, the media, and the public about the risks of germ terrorism and how best to prevent and combat it. That was how we met." See Judith Miller, "Mr. Bio-Defense: William C. Patrick, III: A Tribute," *City Journal*, October 5, 2010, http://www.city-journal.org/2010/eon1005jm.html.

43. Miller, Engelberg, and Broad, *Germs*, 315.

44. The irony that *Germs* unwittingly documents is that the best examples of such migrations of biothreats are to the US biodefense apparatus itself. Alibek and Steven Hatfill, a former Rhodesian paramilitary officer suspected of aiding the white bioterror campaign against Zimbabwean nationalists in the late 1970s, both re-

ceived US state security clearances. On the paramilitary bioterror campaign of the Selous Scouts against Zimbabwe African National Union and the Zimbabwe African People's Union, see Mahvunga, "Vermin Beings."

45. Simon Wessely, "Review: Weapons of Mass Hysteria," *Guardian*, October 19, 2001, http://www.guardian.co.uk/education/2001/oct/20/highereducation.news1. The book highlighted the threat of South Asian–origin "cults" like the Rajneeshees, who poisoned salad bars in Oregon with salmonella in the 1980s, or the Kali-worshipping Japan-based group Aum Shinrikyo, who carried out the sarin chemical attacks on the Tokyo subway in 1995, killing thirteen.

46. On the theory of the Islamic monster terrorist, see Puar and Rai, "Monster, Terrorist, Fag."

47. Miller, Engelberg, and Broad, *Germs*, 109.

48. Alexander Cockburn, "Judy Miller's War," *Counterpunch*, August 16–18, 2003, http://www.counterpunch.org/2003/08/16/judy-miller-s-war/.

49. Judith Miller, "CIA Hunts Iraq Tie to Soviet Smallpox," *New York Times*, December 3, 2002, http://www.nytimes.com/2002/12/03/world/threats-and-responses-germ-weapons-cia-hunts-iraq-tie-to-soviet-smallpox.html.

50. Barton Gellman, "4 Nations Thought to Possess Smallpox," *Washington Post*, November 5, 2002, A1. See also Unmesh Kher, "The Smallpox Scenario," CNN, December 9, 2002, http://www.cnn.com/2002/ALLPOLITICS/12/09/timep.iraq .smallpox2/.

51. Cockburn, "Judy Miller's War"; Iraq Survey Group, *Addendums to the Comprehensive Report of the Special Advisor to the DCI on Iraq's WMD*.

52. Judith Miller, "Verification Is Difficult at Best, Say Experts, and Maybe Impossible," *New York Times*, September 18, 2002, A18.

53. Miller, Engelberg, and Broad, *Germs*, 155–56, 237–38.

54. Libby, *The Apprentice*; Hatfill, "Emergence."

55. In literary theory, the idea of a suspense paradox suggests that suspense generates an affective intensity that is divorced from actual uncertainty. See Carroll, "The Paradox of Suspense."

56. Preston, *The Cobra Event*, xiv.

57. Preston, *The Cobra Event*, xi.

58. Preston, *The Cobra Event*, 111, 115.

59. Masco, "'Sensitive but Unclassified,'" 457; Christopher Marquis, "Powell Blames C.I.A. for Error on Mobile Labs," *New York Times*, April 3, 2004, A5.

60. Drogin, *Curveball*, 81.

61. Masco, "'Sensitive but Unclassified,'" 457.

62. Accounts of the internal deliberations over biosecurity in the United States describe similar displays of vials at policy meetings to demonstrate the apparent ease of deploying weaponized anthrax and other agents. See Miller, Engelberg, and Broad, *Germs*, 215–16; Preston, *Demon in the Freezer*, 221.

63. Lakoff, "The Generic Biothreat," 411.

64. Inglesby, Grossman, and O'Toole, "A Plague on Your City," 437.

65. O'Toole, Mair, and Inglesby, "Shining Light on 'Dark Winter,'" 976.

66. Lakoff, "The Generic Biothreat," 416.

67. UPMC Center for Health Security, "Dark Winter: Findings," accessed August 12, 2015, http://www.upmchealthsecurity.org/our-work/events/2001_dark-winter/about.html.

68. O'Toole, Mair, and Inglesby, "Shining Light on 'Dark Winter,'" 981–82.

69. O'Toole, Mair, and Inglesby, "Shining Light on 'Dark Winter,'" 980–92.

70. Khadra, *The Sirens of Baghdad*, 11, 8, 288, emphasis in original.

71. Hare, *Stuff Happens*.

72. Project for a New American Century, *Iraq*, iii, emphasis added.

73. Project for a New American Century, *Iraq*, ii.

Chapter 5. Refugee Medicine

1. James Dieudonne, quoted in "Haitian Kids Are Mired in Gloom at Naval Base," *Miami Herald*, June 1, 1993, 1A. Jacques Derrida describes the "troubling resemblance" of the beast and the sovereign as an "uncanny hallucination," a ghastly "onto-zoo-anthropo-theologico-political copulation." Derrida, *The Beast and the Sovereign*, 18.

2. Haitian Centers Council, Inc. v. Sale, 92 CV 1258 (SJ) 823 F. Supp. 1028 (1993), 1038.

3. Eric Schmitt, "US Base Is an Oasis to Haitians," *New York Times*, November 28, 1991, http://www.nytimes.com/1991/11/28/world/us-base-is-an-oasis-to-haitians.html.

4. Farmer, "Pestilence and Restraint"; Farmer, "On Guantánamo"; Paik, "Testifying to Rightlessness"; Paik, "Carceral Quarantine at Guantánamo." These essays build on Farmer, *AIDS and Accusation*.

5. Ahmad, "Resisting Guantánamo," 1690.

6. Of course, only a minority of HIV-AIDS patients worldwide have access to such drugs. This is in large part a result of US-led neoliberal patent protections for pharmaceuticals on the international market. See Cooper, *Life as Surplus*, 51–73.

7. Kansas senator Bob Dole stated at the time, "This is not an anti-immigration issue. This is not a gay issue. This is a public health issue, and it is an economic health issue. . . . Even if immigrants enter the country paying their own way, in 10 or 15 years their health may have deteriorated enough to make them a public charge." Dole, quoted in Johnson, "Quarantining HIV-Infected Haitians," 37n142.

8. Kaplan, "Where Is Guantánamo?" Kaplan's important essay, written in the shadow of the Abu Ghraib torture scandal, unfortunately tends to reinforce the Bush administration's use of Guantánamo as a spectacle of US dominance. For a cogent, historically grounded critique of Kaplan via the circuitous routes of imperial exchange, see Braziel, "Haiti, Guantánamo, and the 'One Indispensable Nation.'"

9. Ahmad, "Resisting Guantánamo"; Saar with Novak, *Inside the Wire*.

10. Goldstein, *Storming the Court*, 257–58.

11. Iguanas populate post-9/11 US narratives of Guantánamo in part because

they appear to suggest the greenwashing of the occupation and illegal prison; the camp affords clear protections to endangered nonhuman species but a purposeful ambiguity about the proper treatment of human prisoners. See Saar with Novak, *Inside the Wire*, 42.

12. DeLoughrey, *Routes and Roots*, 40.

13. See the critiques of the racial politics of transmission in Treichler, *How to Have Theory in an Epidemic*; Farmer, AIDS *and Accusation*; Wald, *Contagious*.

14. Farmer, AIDS *and Accusation*.

15. Natalie Gewargis, "Elizabeth Dole Tries to Rename the AIDS Bill after Jesse Helms," ABC News, July 16, 2008, https://web.archive.org/web/20130522013017/http://abcnews.go.com/blogs/politics/2008/07/elizabeth-dole/.

16. Trouillot, "An Unthinkable History."

17. The government even set up special quarantine way stations for Southeast Asian refugees. See Ong, "Making the Biopolitical Subject."

18. Following the refugee crisis at the end of the Vietnam War, the Carter administration used asylum politics to wage the Cold War against avowedly communist states. The United States granted many Southeast Asian refugees asylum, and maintained a policy to declare all Cuban migrants political refugees, famously interdicting boats carrying 100,000 Cubans during the Mariel boatlift of 1980 in order to admit them to the United States rather than to expel them. The United States also admitted 100,000 asylees from Laos, Vietnam, and Cambodia following the Vietnam War. Many of these refugees fled by boat to Hong Kong or overland to Thailand before being brought to the United States and other allied countries.

19. Farmer, AIDS *and Accusation*.

20. Paik, "Testifying to Rightlessness," 44.

21. Henderson, "Surveillance Systems and Intergovernmental Cooperation," 287.

22. American Public Health Association, "Comments on Interim Rule 'Medical Examination of Aliens'" 56 Fed. Reg. 25,000, May 31, 1991, 7; Cheryl Little, "AIDS and Immigration Law," n.d., 272, Americans for Immigrant Justice Records, 1982–2004, Box 17, File: Immigration, Haitians, AIDS 1987–1992, Human Rights Archive, Rubenstein Library, Duke University. All archival documents cited in this chapter are held at the Americans for Immigrant Justice Records at Duke University's Human Rights Archive, henceforth referred to as Americans for Immigrant Justice Records.

23. Duane "Duke" Austin, quoted in *Haitian Centers Council, Inc. v Sale*, 1038.

24. Farmer, "Pestilence and Restraint," 63.

25. Yolande Jean, quoted in Farmer, "Pestilence and Restraint," 61–62.

26. Paik, "Testifying to Rightlessness," 53.

27. Paik, "Testifying to Rightlessness," 52.

28. Wilkins Laguerre, quoted in "Refugee Sings of Hard-Won Freedom," *Miami Herald*, June 17, 1993, 3B.

29. Jean, quoted in Farmer, "Pestilence and Restraint," 52.

30. *Nation*, quoted in Farmer, "Pestilence and Restraint," 56.

31. Florvil Samedi, quoted in "Activist Seeks Justice in Haiti and the US: Former Krome Detainee Fights for Rights," *Miami Herald*, July 8, 1993, 3NE.

32. Fritznel Camy, quoted in Götz-Dietrich, *Haitian Refugees Forced to Return*, 182.

33. See Ahuja, "Abu Zubaydah and the Caterpillar."

34. Paik, "Testifying to Rightlessness," 54.

35. Khanna, "Indignity," 44–45.

36. *USA Today* and the *New England Journal of Medicine*, quoted in Farmer, "Pestilence and Restraint," 64, 62.

37. Bill Clinton, quoted in Paik, "Carceral Quarantine at Guantánamo," 152–53.

38. See the later Supreme Court decision reversing Sterling's order, Sale v. Haitian Centers Council, 509 US 155 (1993).

39. *Haitian Centers Council, Inc. v Sale*, 1038, 1045.

40. Asad, *On Suicide Bombing*, 3.

41. See Dow, *American Gulag*.

42. "INS May Force-Feed Haitians," *Miami Herald*, January 7, 1993, 1A.

43. Letter from Krome detainees, n.d. (presumably January 1 or 2, 1993), relayed by Cheryl Little, Americans for Immigrant Justice Records, 1982–2004, Box 17, File: Articles, Children Stranded in Haiti, 1993, 1 of 3.

44. National Immigration Project of the National Lawyer's Guild, Conference Call Notes, June 24, 1992, Americans for Immigrant Justice Records, 1982–2004, Box 17, File: Immigration, Haitians, AIDS 1987–1992.

45. Goldstein, *Storming the Court*, 255–56, 123; *Sale v. Haitian Centers Council*, 1037.

46. Melinda Cooper claims that autoimmune disorders are a privileged sign of crisis in neoliberal theories of immunity: "Today's immune systems, it would seem, are having trouble distinguishing between the self and the other." See Cooper, *Life as Surplus*, 62.

47. Derrick Z. Jackson, "Ready to Be a Martyr," *Boston Globe*, February 14, 1993, A7.

48. Notable hunger strikes were deployed by Indian nationalists including Mohandas Gandhi, Bhagat Singh, and Jatin Das during the independence struggles of the 1920s–40s and by the Irish Republican Army, which witnessed the starvation death of Bobby Sands in Northern Ireland's infamous Maze prison in 1981. Patrick Anderson points to the highly performative nature of the emaciated body, one that interrupts normative unfoldings of life and death and that transforms subjective temporalities through scenes of self-deprivation. See Anderson, *So Much Wasted*.

49. Jackson, "Ready to Be a Martyr," A7.

50. In this sense, hunger striking is not about transaction, about depriving the body in order to accrue political capital. Its relation to sentiment, tragedy, and spectacle, as well as the eroticization of the vulnerable body, demonstrates the rhetorical complexities of hunger striking and similar, apparently apolitical forms of bodily renunciation such as fasting. See the discussion of the queer spectacle of Mohandas

Gandhi's political "grammar of diet" in Roy, "Meat-Eating, Masculinity, and Re-nunciation in India," esp. 73–74; revised as Roy, *Alimentary Tracts*, chap. 2.

51. Aristide was a prolific liberation theologian and the Louvalas movement often organized around churches and made room for vodou practices within a Hai-tian Catholic framework. Colin Dayan's interpretation of vodou stresses the diffi-culty of reading the syncretic spiritual practices of Haiti within European sacred/secular or sacred/demonic binaries. To serve the lwa, the African spirits, requires specific forms of material sacrifice. Lwa are literally located in the blood (inflecting the attention to blood in fear of HIV infection), and there is an expectation of recip-rocal relations between god and human that lend particular affective investments in sacrifice. See Dayan, *Haiti, History, and the Gods*, esp. 65–70. Farmer's reading of the interpretation of AIDS among rural Haitians stresses the spiritual economy through which infection is traced and fought against. See Farmer, *AIDS and Accusa-tion*, 43–46, 201–7.

52. Elma Verdieu, quoted in Paik, "Testifying to Rightlessness," 56.

53. Yolande Jean, quoted in Paik, "Testifying to Rightlessness," 59.

54. Elma Verdieu, quoted in Paik, "Testifying to Rightlessness," 55.

55. Goldstein, *Storming the Court*, 141.

56. Ratner, "How We Closed the Guantanamo HIV Camp," 208–9; Farmer, "Pestilence and Restraint," 69.

57. Deborah Sontag, "Judge Orders Better Care for Haitians with AIDS," *New York Times*, March 27, 1993, http://www.nytimes.com/1993/03/27/us/judge-orders-better-care-for-haitians-with-aids.html.

58. Lizzy Ratner, "The Legacy of Guantánamo," *Nation*, July 14, 2003, http://www.thenation.com/article/legacy-guantanamo.

59. L. Ratner, "The Legacy of Guantánamo."

60. Department of the Army, *Internment and Resettlement Operations*, Field Manual 3-39.40, February 12, 2010; Detention Hospital, Guantánamo Bay, Cuba, Standard Operating Procedure No. 001, *Voluntary and Voluntary Total Fasting and Re-feeding*, August 11, 2005, http://humanrights.ucdavis.edu/projects/the-guantanamo-testimonials-project/testimonies/testimonies-of-standard-operating-procedures/hunger_strike_sop.pdf.

61. Leonard Doyle, "Inmates' Words: The Poems of Guantánamo," *Independent*, June 21, 2007, http://www.commondreams.org/archive/2007/06/21/2017. "Death Poem" is reproduced in this article.

62. Al-Dossari, "Days of Adverse Hardship in US Detention Camps—Testimony of Guantánamo detainee Jumah al-Dossari" (delivered to Amnesty International), July 2005, https://web.archive.org/web/20090613222232/http://www.amnesty.org/en/library/asset/AMR51/107/2005/en/dom-AMR511072005en.html.

63. Center for Constitutional Rights, "New Hunger Strike Begins after the Department of Defense Reneges on Promises to Detainees," August 31, 2005, http://ccrjustice.org/newsroom/press-releases/new-hunger-strike-begins-after-department-defense-reneges-promises-detainees.

64. Adnan Latif, translated letter to David Remes, December 26, 2010, available from Truthout, http://truth-out.org/images/091512-2-letter.jpg.

65. Glenn Greenwald, "Another Guantánamo Detainee Death Highlights Democrats' Hypocrisy," *Guardian*, September 11, 2012, http://www.guardian.co.uk/commentisfree/2012/sep/11/guantanamo-prisoner-death-democrats; American Civil Liberties Union, "Guantánamo by the Numbers," May 2015, https://www.aclu.org/infographic/guantanamo-numbers.

66. See further Gordon, "Abu Ghraib"; Dayan, *The Law Is a White Dog*.

Epilogue

1. Derrida, *The Animal That Therefore I Am*, 27–29. I refer to Derrida's version of this concept of species war, though the language of a war of species is evident in much older medical and evolutionary discourses. Derrida writes specifically about comparisons of industrial slaughter to the Nazi Holocaust—what he calls a "war waged over the matter of pity." This involves an actual violence by humans against nonhuman species, but also a war over the proper outlook of the human itself. As such, species war is not simply an intractable struggle of groups of organisms with different interests. It is a war over representation, involving battles between humans over the recognition of vulnerable bodies.

2. See Shukin, *Animal Capital*; Vialles, *Animal to Edible*; Pachirat, *Every Twelve Seconds*.

3. Amar, *The Security Archipelago*.

4. Massumi, "National Enterprise Emergency."

5. See US Department of State, *US Government Counterinsurgency Guide*; Matthew et al., *Global Environmental Change and Human Security*.

6. Breyer, quoted in *Department of Health and Human Services v. Florida*, no. 11-398 (March 27, 2012): 63–64.

7. Chakrabarty, "The Climate of History."

8. Ferguson, quoted in Pratt, "Planetary Longings," 210–11.

9. Kaplan, " 'Left Alone with America.' "

10. Ferguson, *Empire*.

11. Ferguson, *Colossus*.

12. McClintock, "Paranoid Empire," 53.

13. Stoler, "Colonial Aphasia," 125. In her article, Stoler writes specifically of the disavowal of French colonialism and racism using the diagnosis of "colonial aphasia," "a dismembering, a difficulty speaking, a difficulty generating a vocabulary that associates appropriate words and concepts with appropriate things."

14. "US capitalism and territorialism were indistinguishable from one another," writes Arrighi (*The Long Twentieth Century*, 59).

15. Desai, *Geopolitical Economy*, 228.

BIBLIOGRAPHY

Abernathy, David. "Bound to Succeed: Science, Territoriality, and the Emergence of Disease Eradication in the Panama Canal Zone." PhD diss., University of Washington, 2000.

Afary, Janet, and Kevin B. Anderson. *Foucault and the Iranian Revolution*. Chicago: University of Chicago Press, 2005.

Agamben, Giorgio. *Homo Sacer: Sovereign Power and Bare Life*. Stanford, CA: Stanford University Press, 1998.

Aguilera-Malta, Demetrio. *Canal Zone*. Tabasco: Editorial Joaquín Mortiz, 1977.

Ahmad, Muneer. "Resisting Guantánamo: Rights at the Brink of Dehumanization." *Northwestern University Law Review* 103, no. 4 (2010): 1683–763.

Ahmed, Sara. "Affective Economies." *Social Text* 79 (2004): 117–39.

Ahuja, Neel. "Abu Zubaydah and the Caterpillar." *Social Text* 106 (2011): 127–49.

Ahuja, Neel. "Macaques and Biomedicine: Notes on Decolonization, Polio, and Changing Representations of Indian Rhesus in the United States, 1930–1960." In *The Macaque Connection: Cooperation and Conflict between Humans and Macaques*, edited by Sindhu Radhakrishna, Michael A. Huffman, and Anindya Sinha, 71–91. New York: Springer, 2013.

Ahuja, Neel. "Notes on Medicine, Culture, and the History of Imported Monkeys in Puerto Rico." In *Centering Animals in Latin American History*, edited by Martha Few and Zeb Tortorici, 180–205. Durham, NC: Duke University Press, 2013.

Ahuja, Neel. "Postcolonial Critique in a Multispecies World." PMLA 124, no. 2 (2009): 556–63.

Alibek, Ken, with Stephen Handleman. *Biohazard: The Chilling True Story of the Largest Covert Biological Weapons Program in the World—Told from the Inside by the Man Who Ran It*. New York: Random House, 1999.

Amar, Paul. *The Security Archipelago: Human-Security States, Sexuality Politics, and the End of Neoliberalism*. Durham, NC: Duke University Press, 2013.

Anderson, Patrick. *So Much Wasted: Hunger, Performance, and the Morbidity of Resistance*. Durham, NC: Duke University Press, 2010.

Anderson, Warwick. *Colonial Pathologies: American Tropical Medicine, Race, and Hygiene in the Philippines*. Durham, NC: Duke University Press, 2006.

Anderson, Warwick. *The Cultivation of Whiteness: Science, Health, and Racial Destiny in Australia*. New York: Basic Books, 2003.

Anderson, Warwick. "Natural Histories of Infectious Disease: Ecological Vision in Twentieth-Century Biomedical Science." *Osiris* 19 (2004): 39–61.

Annan, Kofi. Foreword to *The World Atlas of Great Apes and Their Conservation*, edited by J. Caldecott and L. Miles. Berkeley: University of California Press, 2005.

Appadurai, Arjun. "The Colonial Backdrop." *Afterimage* 24, no. 5 (1997): 4–7.

Appadurai, Arjun. *Fear of Small Numbers: An Essay on the Geography of Anger*. Durham, NC: Duke University Press, 2006.

Arnold, David. *Colonizing the Body: State Medicine and Epidemic Disease in Nineteenth-Century India*. Berkeley: University of California Press, 1993.

Arnold, David, ed. *Imperial Medicine and Indigenous Societies*. Manchester: Manchester University Press, 1988.

Arrighi, Giovanni. *The Long Twentieth Century: Money, Power, and the Origins of Our Times*. London: Verso, 1994.

Arvin, Maile. "Pacifically Possessed: Scientific Production and Native Hawaiian Critique of the 'Almost White' Polynesian Race." PhD diss., University of California–San Diego, 2013.

Asad, Talal. *On Suicide Bombing*. New York: Columbia University Press, 2007.

Ballhatchet, Kenneth. *Race, Sex, and Class under the Raj*. New York: St. Martin's, 1980.

Bartlett, John, Luciana Borio, Lew Radonovich, Julie Samia Mair, Tara O'Toole, Michael Mair, Neil Halsey, Robert Grow, and Thomas V. Inglesby. "Smallpox Vaccination in 2003: Key Information for Clinicians." *Clinical Infectious Diseases* 36 (2003): 883–902.

Battle, Joyce, ed. *Shaking Hands with Saddam Hussein: The US Tilts towards Iraq, 1980–1984*. National Security Archive Briefing Book No. 82, February 23, 2002. http://www.gwu.edu/~nsarchiv/NSAEBB/NSAEBB82/.

Beck, Ulrich. *Risk Society: Towards a New Modernity*. Translated by Mark Ritter. London: Sage, 1992.

Bederman, Gail. *Manliness and Civilization: A Cultural History of Gender and Race in the United States, 1880–1917*. Chicago: University of Chicago Press, 1995.

Bennett, Jane. *Vibrant Matter: A Political Ecology of Things*. Durham, NC: Duke University Press, 2010.

Berlant, Lauren. *Cruel Optimism*. Durham, NC: Duke University Press, 2011.

Berlant, Lauren. "Slow Death (Sovereignty, Obesity, Lateral Agency)." *Critical Inquiry* 33 (2007): 754–80.

Biesanz, John, and Mavis Biesanz. *The People of Panama*. New York: Columbia University Press, 1955.

Boland Osegueda, Roy C. "The Central American Novel." In *Cambridge Companion to the Latin American Novel*, edited by Efraín Kristal, 162–80. Cambridge: Cambridge University Press, 2005.

Brandt, Allan. *No Magic Bullet: A Social History of Venereal Disease in the United States since 1880*. New York: Oxford University Press, 1985.

Braziel, Jana Evans. "Haiti, Guantánamo, and the 'One Indispensable Nation': US Imperialism, 'Apparent States,' and Postcolonial Problematics of Sovereignty." *Cultural Critique* 64 (2006): 127–60.

Breitha, Olivia. *Olivia: My Life of Exile in Kalaupapa*. Honolulu: Arizona Memorial Museum Association, 1988.

Briggs, Charles, with Clara Mantini-Briggs. *Stories in the Time of Cholera*. Berkeley: University of California Press, 2003.

Briggs, Laura. *Reproducing Empire: Race, Sex, Science, and US Imperialism in Puerto Rico*. Berkeley: University of California Press, 2002.

Burnett, Christina Duffy. "The Edges of Empire and the Limits of Sovereignty: American Guano Islands." *American Quarterly* 57, no. 3 (2005): 779–903.

Burt, Jonathan. *Animals in Film*. London: Reaktion, 2002.

Butler, Judith. *Bodies That Matter: On the Discursive Limits of "Sex."* London: Routledge, 1993.

Byrd, Jodi A. *The Transit of Empire: Indigenous Critiques of Colonialism*. Minneapolis: University of Minnesota Press, 2011.

Cabezas, Amalia L. *Economies of Desire: Sex and Tourism in Cuba and the Dominican Republic*. Philadelphia: Temple University Press, 2009.

Carpenter, C. R. "Rhesus Monkeys (*Macaca Mulatta*) for American Laboratories." *Science* 92 (September 27, 1940): 284–86.

Carroll, Noël. "The Paradox of Suspense." In *Beyond Aesthetics: Philosophical Essays*, 254–70. Cambridge: Cambridge University Press, 2001.

Castle, Terry. *The Female Thermometer: Eighteenth-Century Culture and the Invention of the Uncanny*. New York: Oxford University Press, 1995.

Chakrabarty, Dipesh. "The Climate of History: Four Theses." *Critical Inquiry* 35 (2009): 197–222.

Chakrabarty, Dipesh. *Provincializing Europe: Postcolonial Thought and Historical Difference*. Princeton, NJ: Princeton University Press, 2000.

Chamberlain, Winston P. *Twenty-Five Years of American Medical Activity on the Isthmus of Panama, 1904–1929*. Canal Zone: Panama Canal Press, 1929.

Chen, Mel. *Animacies: Biopolitics, Racial Mattering, and Queer Affect*. Durham, NC: Duke University Press, 2012.

Chomsky, Noam. *Rogue States: The Rule of Force in World Affairs*. Boston: South End, 2000.

Chua, Jocelyn Lim. *In Pursuit of the Good Life: Aspiration and Suicide in Globalizing South Asia*. Berkeley: University of California Press, 2014.

Clarke, Adele. *Disciplining Reproduction: Modernity, American Life Sciences, and the Problem of Sex*. Berkeley: University of California Press, 1998.

Clause, Bonnie Treichler. "The Wistar Rat as a Right Choice: Establishing Mammalian Standards and the Ideal of a Standardized Animal." *Journal of the History of Biology* 26 (1993): 329–49.

Coetzee, J. M. *The Lives of Animals*. Princeton, NJ: Princeton University Press, 1999.

Cohen, Ed. *A Body Worth Defending: Immunity, Biopolitics, and the Apotheosis of the Modern Body*. Durham, NC: Duke University Press, 2009.

Cohen, Ed. "The Politics of Viral Containment, or, How Scale Undoes Us One and All." *Social Text* 106 (2011): 15–35.

Colebrook, Claire. "How Queer Can You Go? Theory, Normality, and Normativity." In *Queering the Non/Human*, edited by Noreen Giffney and Myra J. Hird, 17–34. Burlington, VT: Ashgate, 2008.

Commission on the Prevention of WMD Proliferation and Terrorism. *World at Risk: The Report of the Commission on the Prevention of Weapons of Mass Destruction, Proliferation, and Terrorism*. New York: Vintage, 2008.

Committee on Smallpox Vaccination Program Implementation. *The Smallpox Vaccination Program: Public Health in an Age of Terrorism*. Washington, DC: National Academies Press, 2005.

Condé, Maryse. *Tree of Life*. Translated by Victoria Reiter. New York: Ballantine, 1992.

Cooper, Melinda. "The Law of the Household: Foucault, Neoliberalism, and the Iranian Revolution." In *The Government of Life: Foucault, Biopolitics, and Neoliberalism*, edited by Vanessa Lemm and Miguel Vatter, 29–58. New York: Fordham University Press, 2014.

Cooper, Melinda. *Life as Surplus: Biotechnology and Capitalism in the Neoliberal Era*. Seattle: University of Washington Press, 2008.

Cooper, Melinda. "Preempting Emergence: The Biological Turn in the War on Terror." *Theory, Culture, and Society* 23 (2006): 113–35.

Corrales, Scott. *Chupacabras and Other Mysteries*. Murfreesboro, TN: Greenleaf, 1997.

Coughlan, Robert. "Science Moves In on Viruses." *Life*, June 20, 1955, 122–36.

Crosby, Alfred. *The Columbian Exchange: Biological and Cultural Consequences of 1492*. Westport, CT: Greenwood, 1972.

Danner, Mark. "Words in a Time of War: On Rhetoric, Truth, and Power." In *What Orwell Didn't Know: Propaganda and the New Face of American Politics*, edited by Szanto Andras, 16–36. New York: Public Affairs, 2007.

Davis, Lennard. "Introduction: Disability, Normality, and Power." In *The Disability Studies Reader*, 4th ed., edited by Lennard Davis, 1–16. London: Routledge, 2013.

Dayan, Colin. *The Law Is a White Dog: How Legal Rituals Make and Unmake Persons*. Princeton, NJ: Princeton University Press, 2011.

Dayan, Joan. *Haiti, History, and the Gods*. Berkeley: University of California Press, 1995.

Deleuze, Gilles, and Félix Guattari. *A Thousand Plateaus: Schizophrenia and Capitalism*. Translated by Brian Massumi. Minneapolis: University of Minnesota Press, 1987.

DeLoughrey, Elizabeth. *Routes and Roots: Negotiating Caribbean and Pacific Island Literatures*. Honolulu: University of Hawai'i Press, 2009.

Derby, Lauren. "Imperial Secrets: Vampires and Nationhood in Puerto Rico." *Past and Present* 199, supp. 3 (2008): 290–312.

Derrida, Jacques. *The Animal That Therefore I Am*. Translated by David Wills. New York: Fordham University Press, 2008.

Derrida, Jacques. *The Beast and the Sovereign*, vol. 1. Chicago: University of Chicago Press, 2011.

Desai, Radhika. *Geopolitical Economy: After US Hegemony, Globalization, and Empire*. London: Pluto, 2013.

Donnelly, Thomas. *Rebuilding America's Defenses: Strategy, Forces, and Resources for a New Century*. Washington, DC: Project for a New American Century, 2000.

Dow, Mark. *American Gulag: Inside US Immigration Prisons*. Berkeley: University of California Press, 2005.

Drogin, Bob. *Curveball: Spies, Lies, and the Con Man Who Caused a War*. New York: Random House, 2007.

duBois, Page. *Torture and Truth*. London: Routledge, 1991.

Dubos, René. *So Human an Animal*. New York: Scribner, 1968.

Durbach, Nadja. *Bodily Matters: The Anti-vaccination Movement in England, 1853–1907*. Durham, NC: Duke University Press, 2004.

Edmund, Rod. *Leprosy and Empire: A Medical and Cultural History*. Cambridge: Cambridge University Press, 2006.

Eggers, Hans. "Milestones in Early Poliomyelitis Research (1840–1949)." *Journal of Virology* 73, no. 6 (1999): 4533–35.

Esposito, Roberto. *Bíos: Biopolitics and Philosophy*. Translated by Timothy Campbell. Minneapolis: University of Minnesota Press, 2008.

Fanon, Frantz. "Medicine and Colonialism." In *A Dying Colonialism*, translated by Haakon Chevalier, 121–45. New York: Grove, 1965.

Fanon, Frantz. *The Wretched of the Earth*. Translated by Constance Farrington. New York: Grove, 1963.

Fanon, Frantz. *The Wretched of the Earth*. Translated by Richard Philcox. New York: Grove, 2004.

Farmer, Paul. *AIDS and Accusation: Haiti and the Geography of Blame*. 2nd ed. Berkeley: University of California Press, 2006.

Farmer, Paul. "On Guantánamo." In *The Uses of Haiti*, 2nd ed., 217–43. Monroe, ME: Common Cause, 2003.

Farmer, Paul. "Pestilence and Restraint: Guantánamo, AIDS, and the Logic of Quarantine." In *Pathologies of Power*, 51–91. Berkeley: University of California Press, 2003.

Ferguson, Niall. *Colossus: The Rise and Fall of the American Empire*. New York: Penguin, 2005.

Ferguson, Niall. *Empire: The Rise and Demise of the British World Order and Lessons for Global Power*. New York: Basic Books, 2003.

Ferguson, Roderick. "Our Normative Strivings: African American Studies and the Histories of Sexuality." *Social Text* 23 (2005): 85–100.

"First American Monkey Colony Starts on Puerto Rico Islet." *Life*, January 2, 1939, 26–27.

Flamiano, Dolores. "Meaning, Memory, and Misogyny: *Life* Photographer Hansel Mieth's Monkey Portrait." *afterimage* 33, no. 2 (2005): 26.

Flexner, Simon, and Harold L. Amoss. "The Relation of the Meninges and Choroid Plexus to Poliomyelitic Infection." In *Studies from the Rockefeller Institute for Medical Research: Reprints*, vol. 27, 69–82. New York: Rockefeller Institute, 1917.

Forrey, Robert. "Male and Female in London's *The Sea Wolf*." In *Critical Essays on Jack London*, edited by Jacqueline Tavernier-Courbin, 131–40. Boston: G. K. Hall, 1983.

Foucault, Michel. *The Birth of Biopolitics: Lectures at the Collège de France, 1978–1979*. Translated by Graham Burchell. Houndmills, UK: Palgrave, 2008.

Foucault, Michel. "The Birth of Social Medicine." In *The Essential Works of Foucault, 1954–1984*, vol. 3: *Power*, edited by James Faubion, 134–56. New York: New Press, 2000.

Foucault, Michel. *Discipline and Punish*. Translated by Alan Sheridan. New York: Vintage, 1979.

Foucault, Michel. *The History of Sexuality*, vol. 1. Translated by Robert Hurley. New York: Vintage, 1978.

Foucault, Michel. *Security, Territory, Population: Lectures at the Collège de France, 1977–1978*. Translated by Graham Burchell. Houndmills, UK: Palgrave, 2007.

Francione, Gary. *Rain without Thunder: The Ideology of the Animal Rights Movement*. Philadelphia: Temple University Press, 1996.

Franco, Jean. *The Modern Culture of Latin America: Society and the Artist*. Middlesex: Penguin, 1967.

Frazier, Frances N. Introduction to Piʻilani Kaluaikoʻolau, *The True Story of Kaluaikoʻolau as Told by His Wife, Piʻilani*. Lihue, HI: Kauaʻi Historical Society, 2001.

Frederick, Rhonda D. *Colón Man a Come: Mythographies of Panamá Canal Migration*. Lanham, MD: Lexington, 2005.

Frenkel, Stephen. "Geographical Representations of the 'Other': The Landscape of the Panama Canal Zone." *Journal of Historical Geography* 28, no. 1 (2002): 85–99.

Galison, Peter. "Removing Knowledge." *Critical Inquiry* 31, no. 1 (2004): 229–43.

Gandhi, Leela. *Affective Communities: Anticolonial Thought, Fin-de-Siècle Radicalism, and the Politics of Friendship*. Durham, NC: Duke University Press, 2005.

Gandhi, M. K. *Hind Swaraj and Other Writings*. Edited by Anthony J. Parel. Cambridge: Cambridge University Press, 1997.

Gandhi, M. K. *The Story of My Experiments with Truth*. Translated by Mahadev Desai. Ahmedabad, India: Navajivan Press, 1940.

Garland Thomson, Rosemarie. "Seeing the Disabled: Visual Rhetorics of Disability in Popular Photography." In *The New Disability History*, edited by Paul Longmore and Lauri Umansky, 335–74. New York: NYU Press, 2001.

Gerson, Mark, ed. *The Essential Neoconservative Reader*. Reading, MA: Addison Wesley, 1996.

Gikandi, Simon. *Writing in Limbo: Modernism and Caribbean Literature*. Ithaca, NY: Cornell University Press, 1992.

Gilman, Sander. *Disease and Representation: Images of Illness from Madness to AIDS*. Ithaca, NY: Cornell University Press, 1988.

Glick, Megan. "Ocular Anthropomorphisms: Eugenics and Primatology at the Threshold of the 'Almost Human.'" *Social Text* 112 (2012): 97–121.

Goldberg, David Theo. *The Racial State*. Malden, MA: Blackwell, 2001.

Goldstein, Brandt. *Storming the Court: How a Band of Yale Law Students Sued the President—and Won*. New York: Scribner, 2005.

Gonzáles-Martínez, Janis. "The Introduced Free-Ranging Rhesus and Patas Monkey Populations of Southwestern Puerto Rico." *Puerto Rico Health Sciences Journal* 23, no. 1 (2004): 39–46.

Gordon, Avery. "Abu Ghraib: Imprisonment and the War on Terror." *Race and Class* 48, no. 1 (2006): 42–59.

Götz-Dietrich, Opitz. *Haitian Refugees Forced to Return: Transnationalism and State Politics, 1991–1994*. Münster: Lit Verlag, 1999.

Greene, Linda W. *Exile in Paradise: The Isolation of Hawai'i's Leprosy Victims and Development of Kalaupapa Settlement, 1865–Present*. Denver: National Park Service, 1985.

Grosz, Elizabeth. *Volatile Bodies: Toward a Corporeal Feminism*. Bloomington: Indiana University Press, 1994.

Gussow, Zachary. *Leprosy, Racism, and Public Health: Social Policy in Chronic Disease Control*. Boulder, CO: Westview, 1989.

Hage, Ghassan. "'Comes a Time We Are All Enthusiasm': Understanding Palestinian Suicide Bombers in Times of Exigophobia." *Public Culture* 15, no. 1 (2003): 65–89.

Halberstam, Judith. *The Queer Art of Failure*. Durham, NC: Duke University Press, 2011.

Halberstam, Judith. *Skin Shows: Gothic Horror and the Technology of Monsters*. Durham, NC: Duke University Press, 1995.

Haraway, Donna. "A Cyborg Manifesto: Science, Technology, and Socialist-Feminism in the Late Twentieth Century." In *Simians, Cyborgs, and Women: The Reinvention of Nature*, 149–81. London: Routledge, 1991.

Haraway, Donna. *Primate Visions: Gender, Race, and Nature in the World of Modern Science*. London: Routledge, 1989.

Haraway, Donna. *When Species Meet*. Minneapolis: University of Minnesota Press, 2008.

Hardt, Michael, and Antonio Negri. *Empire*. Cambridge, MA: Harvard University Press, 2000.

Hare, David. *Stuff Happens*. New York: Faber and Faber, 2004.

Harlow, Clara Mears. "The Evolution of Harlow Research." In *From Learning to Love: The Selected Papers of H. F. Harlow*, edited by Clara Mears Harlow, xxxiv–xxxv. New York: Praeger, 1986.

Harlow, Harry F. "The Nature of Love." *American Psychologist* 13 (1958): 673–85.

Harvey, David. *Spaces of Capital: Towards a Critical Geography*. London: Routledge, 2001.

Hatfill, Steven. "Emergence." Unpublished manuscript, n.d. Available at the Library of Congress, Washington, DC.

Hayward, Eva. "Lessons from a Starfish." In *Queering the Non/Human*, edited by Noreen Giffney and Myra J. Hird, 249–63. Burlington, VT: Ashgate, 2008.

Henderson, D. A. *Smallpox: The Death of a Disease*. Amherst, NY: Prometheus, 2009.

Henderson, D. A. "Surveillance Systems and Intergovernmental Cooperation." In *Emerging Viruses*, edited by Stephen S. Morse, 283–89. New York: Oxford University Press, 1993.

Henderson, D. A., T. V. Inglesby, J. G. Bartlett, M. S. Ascher, E. Eitzen, P. B. Jahrling, J. Hauer, M. Layton, J. McDade, M. T. Osterholm, et al. "Smallpox as a Biological Weapon." *JAMA* 281, no. 22 (1999): 2127–37.

Herrera, Andres Bolaños. *La Discriminacion de la Mujer en Panamá*. Panamá: Impretex, 1987.

Hird, Myra J. "Animal Trans." In *Queering the Non/Human*, edited by Noreen Giffney and Myra J. Hird, 227–48. Burlington, VT: Ashgate, 2008.

Hitt, A. W. "Leprosy: Its Forms, Characteristics, Distribution, and Treatment." *Medical Standard* 19, no. 9 (1897): 329–33.

Hoganson, Kristin. *Fighting for American Manhood: How Gender Politics Provoked the Spanish-American and Philippine-American Wars*. New Haven, CT: Yale University Press, 1998.

Hua, Julietta. *Trafficking Women's Human Rights*. Minneapolis: University of Minnesota Press, 2011.

Hua, Julietta, and Neel Ahuja. "Chimpanzee Sanctuary: 'Surplus' Life and the Politics of Transspecies Care." *American Quarterly* 65, no. 3 (2013): 619–37.

Hudson, Michael. *Super Imperialism: The Origin and Fundamentals of US World Dominance*. 2nd ed. London: Pluto, 2005.

Hustak, Carla, and Natasha Myers. "Involutionary Momentum: Affective Ecologies and the Sciences of Plant/Insect Encounters." *differences* 23, no. 3 (2012): 74–118.

Inglesby, Thomas, Rita Grossman, and Tara O'Toole. "A Plague on Your City: Observations from TOPOFF." *Clinical Infectious Diseases* 32 (2001): 436–45.

Iraq Survey Group. *Addendums to the Comprehensive Report of the Special Advisor to the DCI on Iraq's WMD*. March 2005. https://www.cia.gov/library/reports/general-reports-1/iraq_wmd_2004/addenda.pdf.

Jameson, Fredric. "Postmodernism, or, the Cultural Logic of Late Capitalism." *New Left Review* 146 (1984): 53–92.

Jerne, Niels. "Nobel Lecture: The Generative Grammar of the Immune System." Lecture given at Karolinska Institutet, Stockholm, December 8, 1984. http://www.nobelprize.org/nobel_prizes/medicine/laureates/1984/jerne-lecture.html.

Jerne, Niels. "Towards a Network Theory of the Immune System." *Annales d'Immunologie* 125C (1974): 373–89.

Johnson, Chalmers. *Blowback: The Costs and Consequences of American Empire.* New York: Macmillan, 2001.

Johnson, Creola. "Quarantining HIV-Infected Haitians: United States' Violations of International Law at Guantanamo Bay." *Howard Law Journal* 37, no. 2 (1994): 305–31.

Jordan, Robert Michael. "El Chupacabra: Icon of Resistance to US Imperialism." Master's thesis, University of Texas–Dallas, 2008.

Kakar, Sanjiv. "Leprosy in British India, 1860–1940: Colonial Politics and Missionary Medicine." *Medical History* 40, no. 2 (1996): 215–30.

Kakar, Sanjiv. "Medical Developments and Patient Unrest in the Leprosy Asylum, 1860 to 1940." In *Health, Medicine, and Empire: Perspectives on Colonial India*, edited by Biswamoy Pati and Mark Harrison, 188–216. New Delhi: Orient Longman, 2001.

Kaluaikoʻolau, Piʻilani. *The True Story of Kaluaikoʻolau as Told by His Wife, Piʻilani.* Līhuʻe, HI: Kauaʻi Historical Society, 2001.

Kaplan, Amy. "'Left Alone with America': The Absence of Empire in the Study of American Culture." In *The Cultures of United States Empire*, edited by Amy Kaplan and Donald Pease, 3–21. Durham, NC: Duke University Press, 1993.

Kaplan, Amy. "Where Is Guantánamo?" *American Quarterly* (2005): 831–58.

Khadra, Yasmina. *The Sirens of Baghdad.* Translated by John Cullen. New York: Anchor, 2007.

Khanna, Ranjana. "Indignity." *positions* 16, no. 1 (2008): 39–77.

Khanna, Ranjana. "Touching, Unbelonging, and the Absence of Affect." *Feminist Theory* 13 (2012): 213–32.

Kim, Clare Jean. "The Racial Triangulation of Asian Americans." *Politics and Society* 27 (1999): 105–38.

King, Nicholas B. "The Scale Politics of Emerging Diseases." *Osiris* 19 (2004): 62–76.

King, Nicholas B. "Security, Disease, Commerce: Ideologies of Postcolonial Global Health." *Social Studies of Science* 32, no. 5–6 (2002): 763–89.

Kohler, Robert. *Lords of the Fly: Drosophila Genetics and Experimental Life.* Chicago: University of Chicago Press, 1994.

Kohn, Eduardo. *How Forests Think: Toward an Anthropology beyond the Human.* Berkeley: University of California Press, 2013.

Kraut, Alan. *Silent Travelers: Germs, Genes, and the "Immigrant Menace."* New York: Basic Books, 1994.

Labor, Earle. "Jack London's Pacific World." In *Critical Essays on Jack London*, edited by Jacqueline Tavernier-Courbin, 205–22. Boston: G. K. Hall, 1983.

Lakoff, Andrew. "The Generic Biothreat, or, How We Became Unprepared." *Cultural Anthropology* 23, no. 3 (2008): 399–428.

Landsteiner, Karl, and Erwin Popper. "Uebertragung der Poliomyelitis acuta auf Affen." *Zeitschrift für Immunitätsforschung, Allergie und Klinische Immunologie* 2 (1909): 377–90.

Lara, Casimiro B. "Leprosy Research in the Philippines: A Historical-Critical Review." Bulletin 10, National Research Council, Philippine Islands, 1936.

Latour, Bruno. "How to Talk about the Body? The Normative Dimension of Science Studies." *Body and Society* 10, no. 2–3 (2004): 205–29.

Lederberg, Joshua. "Appendix C: Testimony of Joshua Lederberg, Ph.D." In *Biological Threats and Terrorism: Assessing the Science and Response Capabilities*, edited by Stacey Knobler, Adef A. F. Mahmoud, and Leslie A. Pray, 235–38. Washington, DC: National Academies Press, 2002.

LeFeber, Walter. *The Panama Canal: The Crisis in Historical Perspective*. New York: Oxford University Press, 1978.

Lewis, Adam. "Naturalizing Empire: Citizenship, Sovereignty, and Antebellum American Literature." PhD diss., University of California–San Diego, 2011.

Leys, Ruth. "The Turn to Affect: A Critique." *Critical Inquiry* 37 (2011): 434–72.

Libby, I. Lewis. *The Apprentice*. New York: St. Martin's, 2005.

Lindsay-Poland, John. *Emperors in the Jungle: The Hidden History of the US in Panama*. Durham, NC: Duke University Press, 2003.

Livingston, Julie, and Jasbir Puar. "Interspecies." *Social Text* 106 (2011): 3–14.

Locke, Constance M. "Peopling an Island with Gibbon Monkeys: An Ambitious West Indian Experiment in Biology." *Illustrated London News*, August 13, 1938, 290.

London, Jack. "Koolau the Leper" [1909]. In *South Sea Tales*, edited by Christopher Gair, 18–33. New York: Modern Library, 2002.

London, Jack. "The Lepers of Molokai." In *Woman's Home Companion* [1908]. http://carl-bell.baylor.edu/JL/TheLepersOfMolokai.html.

Longmore, Paul. "'Heaven's Special Child': The Making of Poster Children." In *The Disability Studies Reader*, 4th ed., edited by Lennard Davis, 34–41. London: Routledge, 2013.

Lowe, Lisa. "The Intimacies of Four Continents." In *Haunted by Empire: Geographies of Intimacy in North American History*, 191–212. Durham, NC: Duke University Press, 2006.

Luhmann, Niklas. *Risk: A Sociological Theory*. Translated by Rhodes Barrett. Berlin: Walter de Gruyter, 1993.

Lundblad, Michael. *Birth of a Jungle: Animality in Progressive-Era US Literature and Culture*. Oxford: Oxford University Press, 2013.

Lye, Colleen. *America's Asia: Racial Form and American Literature, 1893–1945*. Princeton, NJ: Princeton University Press, 2005.

Macfarlane, John T., and Michael Worboys. "The Changing Management of Acute Bronchitis in Britain, 1940–1970: The Impact of Antibiotics." *Medical History* 52, no. 1 (2008): 47–72.

Mahvunga, Clapperton. "Vermin Beings: On Pestiferous Animals and Human Game." *Social Text* 106 (2011): 151–76.

Major, John. *Prize Possession: The United States and the Panama Canal, 1903–1979*. Cambridge: Cambridge University Press, 1993.

Margulis, Lynn, and Dorion Sagan. *What Is Life?* Berkeley: University of California Press, 1995.

Mariner, Wendy, George J. Annas, and Leonard H. Glantz. "*Jacobson v. Massachu-setts*: It's Not Your Great-Grandfather's Public Health Law." *American Journal of Public Health* 95, no. 4 (2005): 581–90.

Markel, Howard. *Quarantine! East European Jewish Immigrants and the New York City Epidemics of 1892*. Baltimore: Johns Hopkins University Press, 1997.

Markel, Howard. *When Germs Travel: Six Major Epidemics That Have Invaded America since 1990 and the Fears They Have Unleashed*. New York: Pantheon, 2004.

Martin, Emily. *Flexible Bodies: Tracking Immunity in American Culture—From the Days of Polio to the Age of AIDS*. Boston: Beacon, 1994.

Marx, Karl. *Capital: A Critique of Political Economy*, vol. 1. Translated by Samuel Moore and Edward Aveling. Moscow: Progress Publishers, 1887.

Masco, Joseph. "Lie Detectors: On Secrets and Hypersecurity in Los Alamos." *Public Culture* 14, no. 3 (2002): 441–67.

Masco, Joseph. "'Sensitive but Unclassified': Secrecy and the Counterterrorist State." *Public Culture* 22, no. 3 (2010): 433–63.

Massumi, Brian. "Fear (the Spectrum Said)." *positions* 13, no. 1 (2005): 31–48.

Massumi, Brian. "National Enterprise Emergency: Steps toward an Ecology of Powers." *Theory, Culture, and Society* 26, no. 6 (2009): 153–85.

Massumi, Brian. *Parables for the Virtual: Movement, Affect, Sensation*. Durham, NC: Duke University Press, 2002.

Matthew, Richard A., Jon Barnett, Bryan McDonald, and Karen L. O'Brian, eds. *Global Environmental Change and Human Security*. Cambridge, MA: MIT Press, 2010.

Mawani, Renisa. "Atmospheric Pressures: On Race and Affect." Unpublished manuscript, n.d.

Mayo, Katherine. *Mother India*. Ann Arbor: University of Michigan Press, 2000.

Mazel, David. "American Literary Environmentalism as Domestic Orientalism." In *The Ecocriticism Reader: Landmarks in Literary Ecology*, edited by Cheryll Glotfelty and Harold Fromm, 137–46. Athens: University of Georgia Press, 1996.

Mbembe, Achille. "Necropolitics." *Public Culture* 15, no. 1 (2003): 11–40.

McAlester, Melani. *Epic Encounters: Culture, Media, and US Interests in the Middle East since 1945*. 2nd ed. Berkeley: University of California Press, 2005.

McAlester, Melani. "Rethinking the 'Clash of Civilizations': American Evangelicals, the Bush Administration, and the Winding Road to the Iraq War." In *Race, Nation, and Empire in American History*, edited by James T. Campbell, Matthew Pratt Guterl, and Robert G. Lee, 352–74. Chapel Hill: University of North Carolina Press, 2007.

McCaffrey, Katherine T. *Military Power and Popular Protest: The US Navy in Vieques, Puerto Rico*. New Brunswick, NJ: Rutgers University Press, 2002.

McClintock, Anne. "Paranoid Empire: Specters from Guantanamo and Abu Ghraib." *Small Axe* 28 (2009): 50–74.

McClintock, James. *White Logic: Jack London's Short Stories*. Grand Rapids: Wolf House, 1975.

McNeill, William H. "Patterns of Disease Emergence in History." In *Emerging Viruses*, edited by Stephen S. Morse, 29–36. Oxford: Oxford University Press, 1996.

McPherson, Alan. "Rioting for Dignity: Masculinity, National Identity, and Anti-US Resistance in Panama." *Gender and History* 19, no. 2 (August 2007): 219–41.

Medevoi, Leerom. "Global Society Must Be Defended: Biopolitics without Boundaries." *Social Text* 91 (2007): 53–79.

Memmi, Albert. *Dependence*. Translated by Phillip A. Facey. Boston: Beacon, 1984.

Mignolo, Walter. *Local Histories/Global Designs: Coloniality, Subaltern Knowledges and Border Thinking*. Princeton, NJ: Princeton University Press, 2000.

Miller, Judith, Stephen Engelberg, and William Broad. *Germs: Biological Weapons and America's Secret War*. New York: Simon and Schuster, 2001.

Mitchell, David, and Sharon Snyder. "Narrative Prosthesis and the Materiality of Metaphor." In *The Disability Studies Reader*, 2nd ed., edited by Lennard Davis, 205–16. London: Routledge, 2006.

Moblo, Pennie. "Blessed Damien of Moloka'i: The Critical Analysis of a Contemporary Myth." *Ethnohistory* 44, no. 4 (1997): 691–726.

Moblo, Pennie. "Defamation by Disease: Leprosy, Myth, and Ideology in Nineteenth Century Hawai'i." PhD diss., University of Hawai'i, 1996.

Moblo, Pennie. "Institutionalizing the Leper: Partisan Politics and the Evolution of Stigma in Post-Monarchy Hawai'i." *Journal of Polynesian Society* 107, no. 3 (1998): 229–62.

Moblo, Pennie. "Review: *The Colony: The Harrowing True Story of the Exiles of Moloka'i* by John Tayman." *Bulletin of the History of Medicine* 81, no. 4 (2007): 889–90.

Molina, Natalia. "Medicalizing the Mexican: Immigration, Race, and Disability in the Early-Twentieth-Century United States." *Radical History Review* 94 (2006): 22–37.

Moran, Michelle. *Colonizing Leprosy: Imperialism and the Politics of Public Health in the United States*. Chapel Hill: University of North Carolina Press, 2007.

Morrow, Prince A. "Leprosy and Hawaiian Annexation." *North American Review*, November 1897.

Mouritz, Alfred. *The Path of the Destroyer: A History of Leprosy in the Hawaiian Islands*. Honolulu: Star-Bulletin, 1916.

Nast, Heidi J., and Michael McIntyre. "Bio(necro)polis: Marx, Surplus Populations, and the Spatial Dialectics of Reproduction and 'Race.'" *Antipode* 43, no. 5 (2011): 1465–88.

Netz, Reviel. *Barbed Wire: An Ecology of Modernity*. Middletown, CT: Wesleyan University Press, 2004.

Ong, Aihwa. "Making the Biopolitical Subject: Cambodian Immigrants, Refugee Medicine, and Cultural Citizenship in California." *Social Science and Medicine* 40, no. 9 (1995): 1243–57.

Oshinsky, David M. *Polio: An American Story*. Oxford: Oxford University Press, 2005.

O'Toole, Tara, Michael Mair, and Thomas V. Inglesby. "Shining Light on 'Dark Winter.'" *Clinical Infectious Diseases* 34 (2002): 972–83.

Pachirat, Timothy. *Every Twelve Seconds: Industrialized Slaughter and the Politics of Sight*. New Haven, CT: Yale University Press, 2011.

Paik, A. Naomi. "Carceral Quarantine at Guantánamo: Legacies of US Imprisonment of Haitian Refugees, 1991–1994." *Radical History Review* 115 (2013): 142–67.

Paik, A. Naomi. "Testifying to Rightlessness: Haitian Refugees Speaking from Guantánamo." *Social Text* 104 (2010): 39–65.

Palmer, Albert. *The Human Side of Hawaii*. Boston: Pilgrim Press, 1924.

Palmer, Steven. "Central American Encounters with Rockefeller Public Health, 1914–1921." In *Close Encounters of Empire*, edited by Gilbert Joseph, Catherine C. LeGrand, and Ricardo D. Salvatore, 311–32. Durham, NC: Duke University Press, 1998.

Panama Canal Department, United States Army. *Control of Venereal Disease and Prostitution in Panama*. Historical Manuscript File, Office of the Chief of Military History, 1947. Microform.

Panama Canal Zone Health Department. "Report." *Social Hygiene* 5, no. 2 (1919): 259–64.

Pandian, Anand. "Pastoral Power in the Postcolony: On the Biopolitics of the Criminal Animal in South India." *Cultural Anthropology* 23, no. 1 (2008): 85–117.

Parascandola, John. "Chaulmoogra Oil and the Treatment of Leprosy." *Pharmaceutical History* 45, no. 2 (2003): 45–57.

Parreñas, Rheana "Juno" Salazar. "Producing Affect: Transnational Volunteerism in a Malaysian Orangutan Rehabilitation Center." *American Ethnologist* 39, no. 4 (2012): 673–87.

Parsons, Talcott. "Sociological Reflections on the United States in Relation to the European War." In *Talcott Parsons on National Socialism*, edited by Uta Gerheart, 189–202. New York: Aldine de Gruyter, 1993.

Pati, Biswamoy, and Mark Harrison, eds. *Health, Medicine, and Empire: Perspectives on Colonial India*. New Delhi: Orient Longman, 2001.

Paul, John R. *A History of Poliomyelitis*. New Haven, CT: Yale University Press, 1971.

Pease, Donald. *The New American Exceptionalism*. Minneapolis: University of Minnesota Press, 2009.

Pennington, Hugh. "Smallpox and Bioterrorism." *Bulletin of the World Health Organization* 81, no. 10 (2003): 762–67.

Phoebus, Eric, Ana Roman, and John Herbert. "The FDA Rhesus Breeding Colony at La Parguera, Puerto Rico." *Puerto Rico Health Sciences Journal* 8, no. 1 (April 1989): 157–58.

Plumwood, Val. *Feminism and the Mastery of Nature*. London: Routledge, 1993.

Podhoretz, Norman. *World War IV: The Long Struggle against Islamofascism*. New York: Doubleday, 2007.

Potter, Will. *Green Is the New Red: An Insider's Account of a Social Movement under Siege*. San Francisco: City Lights, 2011.

Prashad, Vijay. *The Darker Nations: A People's History of the Third World*. New York: New Press, 2008.

Prashad, Vijay. *The Karma of Brown Folk*. Minneapolis: University of Minnesota Press, 2000.

Pratt, Mary Louise. "Planetary Longings." In *World Writing: Poetics, Ethics, Globalization*, edited by Mary Gallagher. Toronto: University of Toronto Press, 2008.

Preston, Richard. *The Cobra Event*. New York: Random House, 1997.

Preston, Richard. *The Demon in the Freezer*. New York: Random House, 2003.

Preston, Richard. Foreword to D. A. Henderson, *Smallpox: The Death of a Disease*, 11–18. New York: Prometheus, 2009.

Project for a New American Century. *Iraq: Setting the Record Straight*. Washington, DC: Project for a New American Century, 2005.

Puar, Jasbir. "Prognosis Time: Toward a Geopolitics of Affect, Debility, and Capacity." *Women and Performance* 19, no. 2 (2009): 161–72.

Puar, Jasbir. *Terrorist Assemblages: Homonationalism in Queer Times*. Durham, NC: Duke University Press, 2007.

Puar, Jasbir, and Ann Pellegrini. "Affect." *Social Text* 100 (2009): 35–38.

Puar, Jasbir, and Amit Rai. "Monster, Terrorist, Fag: The War on Terror and the Production of Docile Patriots." *Social Text* 72 (2002): 117–48.

Pythian, Mark. *Arming Iraq: How the US and Britain Secretly Built Saddam's War Machine*. Boston: Northeastern University Press, 1997.

Rabinow, Paul. "Artificiality and Enlightenment: From Sociobiology to Biosociality." In *Essays on the Anthropology of Reason*, 91–111. Princeton, NJ: Princeton University Press, 1996.

Radford, Benjamin. *Tracking the Chupacabra: The Vampire Beast in Fact, Fiction, and Folklore*. Albuquerque: University of New Mexico Press, 2011.

Ratner, Michael. "How We Closed the Guantanamo HIV Camp: The Intersection of Politics and Litigation." *Harvard Human Rights Journal* 11 (1998): 187–220.

Rawlins, Richard G., and Matt J. Kessler. "The History of the Cayo Santiago Colony." In *The Cayo Santiago Macaques: History, Behavior and Biology*, edited by Richard G. Rawlins and Matt J. Kessler, 13–46. Albany: State University of New York Press, 1986.

Reddy, Chandan. *Freedom with Violence: Race, Sexuality, and the US State*. Durham, NC: Duke University Press, 2011.

Report of Causes Determined in the United States District Court for the District of Hawai'i. Honolulu: Hawaiian Gazette, 1906.

A Report Relating to the Origin and Prevalence of Leprosy in the United States. Senate Document No. 269, 57th Congress, 1st Session, 1902.

Reverby, Susan. *Examining Tuskegee: The Infamous Syphilis Study and Its Legacy*. Chapel Hill: University of North Carolina Press, 2009.

Ritvo, Harriet. *The Platypus and the Mermaid and Other Figments of the Classifying Imagination*. Cambridge, MA: Harvard University Press, 1998.

Rogers, Naomi. *Dirt and Disease: Polio before FDR*. New Brunswick, NJ: Rutgers University Press, 1992.

Rose, Deborah Bird. "What If the Angel of History Were a Dog?" *Cultural Studies Review* 2, no. 1 (2006): 67–77.

Rose, Nikolas. *The Politics of Life Itself: Biomedicine, Power, and Subjectivity in the Twenty-First Century*. Princeton, NJ: Princeton University Press, 2007.

Rosen, Ruth. *The Lost Sisterhood: Prostitution in America, 1900–1918*. Baltimore: Johns Hopkins University Press, 1982.

Rosenberg, Charles. *Explaining Epidemics and Other Studies in the History of Medicine*. Cambridge: Cambridge University Press, 1992.

Roy, Parama. *Alimentary Tracts: Appetites, Aversions, and the Postcolonial*. Durham, NC: Duke University Press, 2010.

Roy, Parama. "Meat-Eating, Masculinity, and Renunciation in India: A Gandhian Grammar of Diet." *Gender and History* 14, no. 1 (2002): 62–91.

Rueckert, William H. "Literature and Ecology: An Experiment in Ecocriticism." In *The Ecocriticism Reader: Landmarks in Literary Ecology*, edited by Cheryll Glotfelty and Harold Fromm, 105–23. Athens: University of Georgia Press, 1996.

Rupke, Nicholaas A. *Vivisection in Historical Perspective*. London: Routledge, 1990.

Ryan, James. *Picturing Empire*. Chicago: University of Chicago Press, 1998.

Saar, Erik, with Viveca Novak. *Inside the Wire: A Military Intelligence Soldier's Eyewitness Account of Life at Guantánamo*. New York: Penguin, 2005.

Saldanha, Arun. *Psychedelic White: Goa Trance and the Viscosity of Race*. Minneapolis: University of Minnesota Press, 2007.

Saldanha, Arun. "Reontologizing Race: The Machinic Geography of Phenotype." *Environment and Planning D: Space and Society* 24 (2006): 9–24.

Salvatore, Ricardo. "Imperial Mechanics: South America's Hemispheric Integration in the Machine Age." *American Quarterly* 58, no. 3 (2006): 663–91.

Schelling, Thomas. *The Diplomacy of Violence*. New Haven, CT: Yale University Press, 1966.

Schiebinger, Londa. *Nature's Body: Gender in the Making of Modern Science*. Boston: Beacon, 1993.

Scott, James C. *Seeing Like a State: How Certain Schemes to Improve the Human Condition Have Failed*. New Haven, CT: Yale University Press, 1998.

Seabrook, William. *The Magic Island*. New York: Harcourt, 1929.

Segelid, Michael J. "Smallpox Revisited." *American Journal of Bioethics* 3, no. 1 (2003): W5–W11.

Serlin, David. *Replaceable You: Engineering the Body in Postwar America*. Chicago: University of Chicago Press, 2004.

Serra, Yolanda Marco. "El Movimiento Sufragista en Panamá y la Construcción de la Mujer Moderna." In F. Aparicio, et al., *Historia de los Movimientos de Mujeres en Panamá en el Siglo XX*, 45–132. Panamá: Instituto de la Mujer de la Universidad de Panamá, 2002.

Shah, Nayan. *Contagious Divides: Epidemics and Race in San Francisco's Chinatown*. Berkeley: University of California Press, 2001.

Shah, Svati. "Prostitution, Sex Work, and Violence: Discursive and Political Contexts for Five Texts on Paid Sex, 1987–2001." *Gender and History* 16, no. 3 (2004): 794–812.

Shakespeare, Tom. "The Social Model of Disability." In *The Disability Studies Reader*, 2nd ed., edited by Lennard Davis, 197–204. London: Routledge, 2006.

Shell, Marc. *Polio and Its Aftermath: The Paralysis of Culture*. Cambridge, MA: Harvard University Press, 2005.

Shukin, Nicole. *Animal Capital: Rendering Life in Biopolitical Times*. Minneapolis: University of Minnesota Press, 2009.

Silva, Denise Ferreira da. *Toward a Global Idea of Race*. Minneapolis: University of Minnesota Press, 2007.

Silva, Noenoe K. *Aloha Betrayed: Native Hawaiian Resistance to American Colonialism*. Durham, NC: Duke University Press, 2005.

Silva, Noenoe K., and Pualeilani Fernandez. "Mai Ka ʻĀina O Ka ʻEhaʻeha Mai: Testimony of Hansen's Disease Patients in Hawaiʻi, 1866–1897." *Hawaiian Journal of History* 40 (2006): 75–97.

Sinán, Rogelio [Bernardo Domínguez Alba]. *Plenilunio*. Panamá: Impresora Panamá, S.A., 1961.

Siu, Lok. *Memories of a Future Home: Diasporic Citizenship of Chinese in Panama*. Stanford, CA: Stanford University Press, 2005.

Spivak, Gayatri Chakravorty. "Can the Subaltern Speak?" In *Colonial Discourse and Post-colonial Theory*, edited by Patrick Williams and Laura Chrisman, 64–111. New York: Columbia University Press, 1994.

Spivak, Gayatri Chakravorty. *A Critique of Postcolonial Reason: Toward a History of the Vanishing Present*. Cambridge, MA: Harvard University Press, 1998.

Spivak, Gayatri Chakravorty. *Death of a Discipline*. New York: Columbia University Press, 2003.

Stepan, Nancy. *Eradication: Ridding the World of Diseases Forever?* Ithaca, NY: Cornell University Press, 2011.

Stepan, Nancy. *Picturing Tropical Nature*. Ithaca, NY: Cornell University Press, 2001.

Stern, Alexandra Minna. *Eugenic Nation: Faults and Frontiers of Better Breeding in Modern America*. Berkeley: University of California Press, 2005.

Stern, Alexandra Minna. "The Public Health Service in the Panama Canal: A Forgotten Chapter of US Public Health." *Public Health Reports* 120 (2005): 675–80.

Stevenson, Robert Louis. *Father Damien: An Open Letter to the Reverend Dr. Hyde of Honolulu* [1889–1890]. Boston: Alfred Bartlett, 1900.

Stoddard, Charles Warren. *The Lepers of Molokai* [1885]. Notre Dame, IN: Ave Maria Press, 1908.

Stoler, Ann Laura. "Colonial Aphasia: Race and Disabled Histories in France." *Public Culture* 23, no. 1 (2011): 121–56.

Stoler, Ann Laura. "Intimidations of Empire: Predicaments of the Tactile and Unseen." In *Haunted by Empire: Geographies of Intimacy in North American History*, 1–22. Durham, NC: Duke University Press, 2006.

Stoler, Ann Laura. *Race and the Education of Desire: Foucault's* History of Sexuality *and the Colonial Order of Things.* Durham, NC: Duke University Press, 2005.

Streeby, Shelley. *American Sensations: Class, Empire, and the Production of Popular Culture.* Berkeley: University of California Press, 2002.

Subcommittee on Pacific Islands and Porto Rico. *Hawaiian Investigation Report.* Washington, DC: Government Printing Office, 1903.

Sunder Rajan, Kaushik. *Biocapital: The Constitution of Postgenomic Life.* Durham, NC: Duke University Press, 2006.

Suskind, Ron. "Faith, Certainty, and the Presidency of George Bush." *New York Times Magazine,* October 17, 2004. http://www.nytimes.com/2004/10/17/magazine/17BUSH.html.

Tayman, John. *The Colony.* New York: Scribner, 2006.

Tenet, George. *At the Center of the Storm: The CIA during America's Time of Crisis.* New York: HarperCollins, 2007.

Thornton, Russel. *American Indian Holocaust and Survival: A Population History since 1492.* Norman: University of Oklahoma Press, 1987.

Thrailkill, Jane. *Affecting Fictions: Mind, Body, and Emotion in American Literary Realism.* Cambridge, MA: Harvard University Press, 2007.

Thrift, Nigel. *Non-representational Theory: Space, Politics, Affect.* London: Routledge, 2008.

Tomes, Nancy. *The Gospel of Germs: Men, Women, and the Microbe in American Life.* Cambridge, MA: Harvard University Press, 1999.

Tomso, Gregory. "The Queer History of Leprosy and Same-Sex Love." *American Literary History* 14, no. 4 (winter 2002): 747–75.

Treichler, Paula. *How to Have Theory in an Epidemic: Cultural Chronicles of AIDS.* Durham, NC: Duke University Press, 1999.

Trouillot, Michel-Ralph. "An Unthinkable History." In *Silencing the Past: Power and the Production of History,* 70–107. Boston: Beacon, 1995.

Tucker, Jonathan. *Scourge: The Once and Future Threat of Smallpox.* New York: Grove, 2002.

Turse, Nicolas. "Experimental Dreams, Ethical Nightmares: Leprosy, Isolation, and Human Experimentation in Nineteenth Century Hawai'i." In *Imagining Our Americas: Toward a Transnational Frame,* edited by Sandhya Shukla and Heidi Tinsman, 138–67. Durham, NC: Duke University Press, 2007.

Uexküll, Jakob von. *A Foray into the Worlds of Animals and Humans with a Theory of Meaning.* Translated by Joseph D. O'Neill. Minneapolis: University of Minnesota Press, 2010.

Unruh, Vicky. *Latin American Vanguards: The Art of Contentious Encounters.* Berkeley: University of California Press, 1994.

US Department of State. *US Government Counterinsurgency Guide.* Washington, DC: Bureau of Political-Military Affairs, 2009.

Vandenbergh, John G. "The La Parguera, Puerto Rico Colony: Establishment and Early Studies." *Puerto Rico Health Sciences Journal* 8, no. 1 (April 1989): 117–19.

Vaughan, Meghan. *Curing Their Ills: Colonial Power and African Illness*. Stanford, CA: Stanford University Press, 1991.

Vialles, Noilie. *Animal to Edible*. Cambridge: Cambridge University Press, 1994.

Virilio, Paul. *Desert Screen: War at the Speed of Light*. Translated by Michael Degener. London: Continuum, 2002.

Wailoo, Keith. *Dying in the City of the Blues*. Chapel Hill: University of North Carolina Press, 2001.

Wald, Priscilla. *Contagious: Cultures, Carriers, and the Outbreak Narrative*. Durham, NC: Duke University Press, 2007.

Walrond, Eric. "Godless City." In *"Winds Can Wake Up the Dead": An Eric Walrond Reader*, edited by Louis J. Parascandola, 161–72. Detroit: Wayne State University Press, 1998.

Walrond, Eric. "Morning in Colón." In *"Winds Can Wake Up the Dead": An Eric Walrond Reader*, edited by Louis J. Parascandola, 310–14. Detroit: Wayne State University Press, 1998.

Weatherall, M. *In Search of a Cure: A History of Pharmaceutical Discovery*. Oxford: Oxford University Press, 1991.

Weheliye, Alexander. *Habeas Viscus: Racializing Assemblages, Biopolitics, and Black Feminist Theories of the Human*. Durham, NC: Duke University Press, 2014.

Weltzin, Richard, Jian Liu, Konstantin V. Pugachev, Gwendolyn A. Myers, Brie Coughlin, and Paul S. Blum. "Clonal Vaccinia Virus Grown in Cell Culture as a New Smallpox Vaccine." *Nature Medicine* 9 (2003): 1125–30.

Wilson, Daniel. "A Crippling Fear: Experiencing Polio in the Era of FDR." *Bulletin of the History of Medicine* 72, no. 3 (1998): 464–95.

Wolfe, Cary. *Critical Environments: Postmodern Theory and the Pragmatics of the "Outside."* Minneapolis: University of Minnesota Press, 1998.

Worms, Werner. "Prophylaxis of Syphilis by Locally Applied Chemicals: Methods of Examination, Results, and Suggestions for Further Experimental Research." *British Journal of Venereal Diseases* 16, no. 3–4 (1940): 186–210.

INDEX

austerity, 199
azidothymidine (AZT), 175

Bacillus leprae. See Hansen's disease
Bateson, Gregory, xv
Bazin, Marc Louis, 184
Bedtime for Bonzo, 125
behavioralism, 16
Beleño, Joaquin, 95
Belgian Congo, 114
Berlant, Lauren, xi
bin Laden, Osama, 4, 200
Biological and Chemical Weapons Convention, 146
biomedicine: anxieties over, 117; portrayal in film, 118
bio-necro collaboration, xi–xii, 108, 202
biopolitics, vii–xiii, ix, xi–xiii, 26–27, 67–69
bioprospectors, 125
biosecurity, 15, 22, 152; and animal research, 131; anticipatory approach to, 167; expansionary regime of, 24; measures following 9/11, 3; military, 135. *See also* security
biosemiotics, 15
bioterrorism, 138; anti-imperialist fantasy of, 165–66; campaign against Zimbabwean nationalists, 223n44; fear produced by exercises, 162; response teams, 155; scenario-based exercises, 158
bioweapons, 3–4, 152; in French and Indian Wars, 142–43; mobile laboratories, 156–57; research, 145
birth control, 179
Blue Moon Girls, 86–87
boat people, Haitian, 176–77
bodies: anthropomorphized, 181; controlled by empire, 4–8; critical approaches to, 9; depictions of black, 171; dismembered, 50, 56; effects of hunger striking on, 187, 227n48; horror at animalized and debilitated, 56; hybrid, 41, 119–20; incarcerated, 181; materiality of, 5; nonconsensual appropriation of, 124; objects of governance, vii; panic aimed at men's, 77; plasticity of, 140; politics of, 25–26, 65; primate, xv;

public fears and hopes about, viii; quarantine of women's, 78–79; racial suspicion of, 151–52; of refugees, 183, 189; resistance of colonized, 25, 186, 190–91, 201; of rhesus macaques, 108–9; of settlers, 14; sexual urges of, 86; site for restoring family, 118; transitions of, 8–11, 47; vulnerable, vii
Boston, smallpox epidemic of 1901–3, 1–2, 4
Bouquet, Henry, 142
Bracken, H. M., 37
Brandt, Allan, 81
Breitha, Olivia, 32, 62–64
Breyer, Stephen, 198
Briggs, Laura, 77
Broad, William, 152, 160
Brodie, Maurice, 109
brothels: colonized countries portrayed as, 94; sites of covert resistance, 98; soldiers banned from, 81
Buck v. Bell, 209n2
Burch, George A., 113, 115, 130
Burnet, F. Macfarlane, 13–14
Bush, George H. W., 170
Bush, George W., 136, 138, 166–67
Bush doctrine, 166–67
Byrd, Jodi A., 143

cabarets, 86, 95–98
Cabezas, Amalia, 87
camp, 65, 68–69; AIDS concentration camp, 170, 182; redrawing borders of species, 179–80
Canal Zone (Aguilera-Malta), 96
Canóvanas, Puerto Rico, 127
Cape Colony, South Africa, 40, 43
capital, limits to, 207n5
Capital (Marx), 207n5
capitalism, xiv, 32, 177
Caribbean modernism, 218n50
Caribbean Regional Primate Center, 220n28
Carpenter, Clarence Ray, 109, 111–12
carriers, asymptomatic, 12, 20, 107
Carter, Jimmy, 177, 226n18
Carville National Leprosarium (Louisiana), 46, 62, 211n34

Cayo Santiago, Puerto Rico, 21, 101–4, 109, 111–12, 220n28

CDC, 20, 143, 145, 175–76

Cédras, Raoul, 184

Centers for Disease Control (CDC), 20, 143, 145, 175–76

Central Intelligence Agency (CIA), 4, 138, 157, 200

chaulmoogra oil, 66–67

chemical prophylaxis, 87

Chen, Mel, 8

Cheney, Dick, 3, 147

chimpanzees, 121, 125; field stations for capture and breeding of, 114; nuclear weapons research on, 112; photographs of, 103. *See also* primates

Chinese immigrants, association with Hansen's disease, 35, 37–38, 50, 53, 55

chupacabras, 127–28, 221n47

CIA, 4, 138, 157, 200

clinical trials, 20–21, 108

Clinton, Bill, 138, 145–49, 155, 182, 186–89

Coast Guard, US, 169–70

Cobra Event, The (Preston), 155–56

Cockburn, Alexander, 153

Cohen, Ed, 8

Cohen, William, 167

Cold War: competition with Soviet Union, 20; deterrence during, 148; ideologies, 177; politics of decolonization and containment, 21

Colón, Panamá, 75, 81, 91

"Colonial Aphasia" (Stoler), 229n13

colonialism: architectures of control, 68; and disease ecology, 13; enclosure, 21; resistance to, 25–26

Colony, The (Tayman), 213n13

Columbia School of Tropical Medicine, 111

Committee for the Establishment of a National Cardiovascular Primate Colony at NIH, 113

Compagnie de Francaise de Navigation a Vapeur v. Louisiana State Board of Health, 18

compulsory treatment, 2, 19, 81

Condé, Maryse, 94

consent, informed, 65–67, 69, 179

contagions: association with animals, 10; black revolt as, 176; fears of, 9–10; movement of, 2–3; and settler claims to sovereignty, 48; and sexuality, 38

containment: anticipatory, 15; Cold War politics of, 21; of communism and decolonization, 104; of Haitian immigrants, 172–73; of interspecies contact, 195; nuclear, 139; spatial techniques of, 7; of surplus populations, xiv; of venereal disease, 87–88; war as method of, 148

Control of Venereal Disease and Prostitution in Panama (US Caribbean Command), 80

Cooper, Melinda, xiv, 137–38

Costa Rica, US-sponsored public health efforts, 215n2

counter-conduct, 26

criminalization: of Hansen's disease patients, 33; of women in urban space, 73, 79

crisis thinking, 22–23, 199, 202

Crosby, Alfred, 222n13

cross-cultural encounters, 87

cross-dressing, 59–61

Cuba, 75

Cueva, Puerto Rico, 126

Cultures of United States Imperialism, The (Kaplan and Pease), 202

Danzig, Richard, 155, 158

Darjani Primate Center, Kenya, 115

"Dark Biology" trilogy (Preston), 155

Dark Winter exercise, 160–63

Darwin, Charles, 207n5

da Silva, Denise Ferreira, 58

Dayan, Colin, 228n51

"Death Poem" (al-Dossari), 190–91

deathworlds, 69

decolonization, 20

Deleuze, Gilles, xv

democracy: evangelism of, 203; US defense of, vii

Demon in the Freezer, The (Preston), 155

de Paredes, Guillermó Garcia, 92

Department of Natural Resources and Environment (DRNA), Puerto Rico, 129
Dependence (Memmi), 215n54
dependency, 215n54; of blackness, 51; colonial, 48–59; Haiti as model of, 176; racialized, 199; refusal of, 57; sexuality as form of, 58
Derby, Lauren, 127
Derrida, Jacques, 195–96, 229n1
Desai, Radhika, 204
destigmatization theory, 211n34
detention, indefinite, 172, 179–80, 183, 190
deterrence, 148
development, undesirable consequences of, 14
Dieudonne, James, 170
disability, 122–24; disability studies, 9; metaphor for social disempowerment, 57
discipline, military, 86, 174–75
disease interventions, 4–5; biopolitical critique of, xiii; catastrophe as justification for, 199; challenges to, 24, 200; controversies over, 2; defense of national body, 3; expansion of, 196; fears of, 21; imperial, 147–48; logics of, 201; militarized, 168; as political processes, 195; unilateral, 149; use of state authority and, 7
diseases: at borders of empire, 179; control of, 11; drug-resistant, 22; as effect of settlement and trade, 35; eradication model of, 21, 72; global surveillance system, 178–79; insect-borne, 75; invisible transmission of, 12; neoliberal conceptions of, 137, 139; new outbreaks of, 22; racialization of, 5–6, 13–15
disorder: attributed to Haiti, 176; fear of, 136
dread life, xi, 6, 9–10, 130, 197, 203
dual-use research, 136, 144–46, 150
Dubos, René, 13–14
Duvalier, Jean-Claude "Baby Doc," 177

Ebola, 154–55
ecologies: alteration of, 75; of disease, 12; of incarceration, 181; sexual, 76–79
Edenic Polynesian effect, 51

Eighth Amendment protections for incarcerated, 185
elephantiasis (lymphatic filariasis), 41–43
El Panamá América, 91–92
El Panamá Estrella, 89–90
El Salvador, US-sponsored public health efforts, 215n2
emergence, language of, 147–48
emergency powers, 7–8
emerging diseases, 22, 137–39, 178–79
Emerging Viruses (Lederberg), 142
empire: biopolitics of, vii; changing articulations of, 205; class-based critique, 55; and control of bodies, 4–8; delimitation of sovereignty, 69–70; disease risk at borders of, 179; historians of, 204; intervention into domains of life, vii; management of affective relations, xi; management of life and death, x; Marxist theories of US, 204, 211n35; masculine resistance to, 54–55; myths of benevolent, 19; paranoia of, 203; as project in government of species, x; psychologizing of imperialism, 203; regenerative forms of living within, 26–27; studies of, viii–ix. *See also* United States empire
Engleberg, Stephen, 152
environmental crises, studies of, 201
environmentality, 198
environmental movements, xiv
environmental warfare, 72; challenges to, 17
eradication model of disease control, 21, 72–75
evolutionary theory, 207n5
exchange, biosocial forms of, x
exclusion: of HIV-positive immigrants, 173, 175–76; militarized, 171; space of, 65–67
exoticism, xv, 46–47. *See also* orientalism
experiments: debate over, 67; on humans, 30, 65–67, 108, 120; on polio, 107–8; radiological, 112; and slavery, 69; on smallpox, 145–46

fallen city trope, 93–99
Fanon, Frantz, 25, 93–94, 201

Farmer, Paul, 169–71, 175, 181
FDA, 126
fear: of biomedical intervention, 21, 118–19; of disease transmission, 9; management of, 15; produced by bioterrorism exercises, 162
feminists, Panamanian, 76
feminist studies, xiii–xiv, 9
feminization, 58
Fenner, Frank, 13
Ferguson, James, 202
Ferguson, Niall, 203
Ferguson, Roderick, 81
film, 116–26
fire, symbolism of, 97–98
Fitch, George, 66–67
Flexner, Simon, 106–7
Food and Drug Administration (FDA), 126
force-feeding, 191
force projection, 148
Foucault, Michel, xi–xiii, 26, 67–69
French and Indian Wars, 142–43
Frenkel, Stephen, 74
Freud, Sigmund, 212n41
Fukuyama, Francis, 147

Galison, Peter, 150
Gandhi, M. K., 25–26, 110
Garland Thomson, Rosemarie, 43, 46–47
gaze: blurring of sexual and medical, 53; distinct from stare, 43; eroticized, 55–56; medical, 46–47, 62; photographic, 102–3
Gellman, Barton, 154
gender: bending of roles, 59–62; and nationalist anti-imperialism, 94; and venereal disease, 72. *See also* feminization; masculinity
Geneva Conventions, 191
Gergen, David, 160
Germs (Miller, Broad, and Engleberg), 152–58
germ theories, 30, 44–46
Gibson, Walter, 48–49
Gikandi, Simon, 218n50
Gilman, Sander, 40

Ginsburg, Ruth Bader, 169
Glick, Megan, 103, 121
"Godless City" (Walrond), 97
Gold and Silver Roll system, 75
Gorgas, William, 75
gorillas, 118–22. *See also* primates
gothic imagery, 10, 15, 40, 52–57, 80–81, 180–81
government of species: empire as project of, x, 10–11; in film, 124; history of, 196; managing epidemics, 17–19; and posthuman idealism, xv–xvi; race and, 121–22, 197; structural and affective dimensions of, 167; unruliness of, 24, 26
Grant, Cary, 125
Guantánamo, Cuba: architecture of, 174, 180; post-9/11 detentions, 23, 189–93; press tours of, 192–93; prisoner abuse at, 191; public relations representations of, 192
Guattari, Félix, xv
Guayacán, Puerto Rico, 126
Guayaquil, Ecuador, 96
Gussow, Zachary, 38

Habeas Viscus (Weheliye), 208n15
Haitian Centers Council, 184, 186
Haitian Centers Council, Inc. v. Sale, 23
Haitian refugees, 169–89; housed in US jails, 184; interdiction at sea, 178; lawsuit on behalf of, 184–89; refused legal rights, 177–78; US policy toward, 177
Handbook of Leprosy, A (Impey), 40
Hansen, Armauer, 66
Hansen's disease, 18–19; adoption of term, 211n34; and annexation of Hawai'i, 32; appearance in Hawai'i, 33; association of ethnic groups with, 35–38, 50–55; Britain's failure to eliminate, 40; British discourse on, 213n19; colonial history of, 64; current medical understanding, 38; discovery of bacterium, 28; end of forced segregation for, 62; eroticized images of, 53–54; experimental treatments for, 66; geographies of, 32–38; gothic images of,

Israel, 147

Isthmanian Canal Commission, 75

Jacobson, Henning, 1–3

Jacobson v. Massachusetts, 1–2, 11–12, 20, 34, 36, 198

Jahrling, Peter, 146

James, C. L. R., 175

Jean, Yolande, 169–70, 180, 187–88

Jenner, Edward, 142

Jerne, Niels, 12–13

Jim Crow segregation, 74–75

Johnson, Sterling, 178, 183, 186, 188–89

Johnson Atoll, 145, 156

Jordan, Robert Michael, 128

jungle, film depictions of, 125

Jungle Jim, 125

Jungle Queen, 125

Justice Department, US, 190

Kaipu, Mikala, 34

Kalalau valley, Hawai'i, 54, 61

Kalihi Leprosy Hospital, 47, 62–64

Kaluaiko'olau (Koolau), 32, 54–59

Kaluaiko'olau, Pi'ilani, 59–62

Kamehameha, Lot, 33

Kaplan, Amy, 174, 202

Karloff, Boris, 117–18

Kazakhstan, 154

Kerry, John, 149

Khadra, Yasmina (Mohammed Moulessehoul), 165

Khalilzad, Zalmay, 147

Khanna, Ranjana, 182

kinesis, 122–24

Kipling, Rudyard, 203

knowledge: state monopoly on, 150; and threat of bioweapons, 152

Kolmer, John, 109

Koltsovo, Siberia, 143

Koolau, 32, 54–59

"Koolau the Leper" (London), 54–59

Korematsu v. United States, 209n2

Kristol, William, 147

Krome immigration jail, 184

laboratory: film depictions of, 125; securitization of, 131; smallpox stocks stored in, 144

Laboratory of Perinatal Psychology, 126

labor markets, 51

labor movements, 50

Lacy, Rachel, 209n3

La Estrella de Panamá, 89–90

LaGuerre, Wilkins, 181

Lajas Valley, Puerto Rico, 126, 128

Lakoff, George, 15, 137, 160–61

Landsteiner, Karl, 106–7

La Parguera, Puerto Rico, 126

lateral agency, xi

Latif, Adnan, 191

Lederberg, Joshua, 146–48, 155

Lee, Wen Ho, 151

"Lepers of Molokai, The" (London), 50, 52, 55

Lepers of Molokai, The (Palmer), 35

leprosy. *See* Hansen's disease

Levaditi, Constantin, 106–7

Lewis, Paul A., 107

Libby, I. Lewis "Scooter," 145, 147, 152–53, 155

life: inability of state to control, 201; instrumentalism of, 182; interconnected, 8; power of medical science over, 119; rendered into capital, 196; role in imperial domination, 205; state and corporate efforts to control, 198; technocratic power over, 24; zone between life and death, 65

Life magazine, 101, 105, 110

Lindsay-Poland, John, 75

literary leakers, 135–36, 155–56

London, Jack, 49–59

Louisiana, 18, 32, 36, 211n34

Lowe, Lisa, xiii, 130

Lugosi, Bela, 119–20

Luhmann, Niklas, xv

lumbar puncture, 106, 119

Luna Verde (Beleño), 95

Lundblad, Michael, 121

Lye, Colleen, 50–51, 59

nature: alienation from, vii; corruption of, 117; Hawaiians' relation to, 55; threat to medical modernity, 121

necropolitical theories, xii, 19, 31–32, 69

"Necropolitics" (Mbembe), 69

Nehru, Jawaharlal, 113

neoconservatism, 136, 139, 147, 166, 196

neoliberalism, xii, 196–97

nervous system, connection to immune and digestive systems, 16–17

networked immunity, 12

New Deal Civilian Conservation Corps, 111

new materialism, viii, xv, 207n5

newspapers: Haitian refugees' hunger strikes reported in, 186; reports of anti-vice activities in Panamá, 89–90

New Yorker, 154–55

New York Times, 152–53, 170, 181

Nigh, William, 117

NIH (National Institutes of Health), 20–21; primate centers, 129–31; promotion of term "primate," 125

Nixon, Richard, 152

nonhuman: agency of, viii; use of term, 130–31

nonviolence, 26

normalization, of public health measures, 6

North American Review, 36

Nunn, Sam, 160–61

Obama, Barack, 176

Obeah, 98

Olivia (Breitha), 62–64

opportunistic infections (AIDS), 185

optimism: about imperial intervention, 6, 9; about medical advances, 104–5; about technology, 14, 117

oral pathway. *See* mouth

orientalism, xv, 53, 56, 101, 110. *See also* exoticism

otherness, romantic notions of, xv–xvi

O'Toole, Tara, 160

Outbreak (film), 155

outbreak narratives, 9–10, 154–55

Pacific islander (racial category), 59

Paik, Naomi, 171, 177, 180, 188

Pakistan, 4

Palmer, Albert, 35

Panama Canal Department of the Army, 77

Panamá Canal Zone, 18–19, 71–99; abundance of single men, 92; borders of, 74; feminization of, 96; incarceration of sex workers, 19; influx of immigrants, 78, 83–84; juridical structure of, 75; legal status of residents, 76; racial apartheid in, 216n5

Panamá City, 75, 81, 95

Panama Tribune, The, 90

Panamerican Sanitary Conference, 92

panics: aimed at men's bodies, 77; over migration, 199

paralysis, 118–24

paranoia, 203

paranormal activity, 127

Park, William, 109

Parsons, Talcott, xv

Path of the Destroyer, The (Mouritz), 28

Pathologies of Power (Farmer), 169

patient activists, 19, 32, 62, 64–65, 200

Patrick, William, III, 145, 152, 160

Pease, Donald, 202

penicillin, 90, 104–5

Perle, Richard, 147

pharmaceuticals, 20–21, 90

philanthropists, 52, 215n2

Philippines, 34

photography, 101–4; of chimpanzees, 103; colonial, 47; of Hansen's disease, 38–48, 62–64; political character of, 40; of small-pox patients, 163–64

Plame, Valerie, 152

plantations, 55

Plenilunio (Sinán), 95

Podhoretz, Norman, 147–48, 223n26

polio, 20–21; asymptomatic carriers, 107; clinical trials of vaccines, 108; development of vaccine, 105–6; experiments on, 107–8; fears associated with, 104; polio

polio (*continued*)

scare of 1948–55, 106, 112; rhesus macaques used in research, 103; working class and immigrants targeted in control efforts, 106

Popper, Edwin, 107

port quarantines, 3, 46

postcolonial studies, ix, xiii, 17, 197, 201–2, 205

posthumanism, viii–xvi

Powell, Colin, 156–58, 159, 166

power: doubleness of, 203; and reshaping of life, xi; transborder and transpecies routes of, 17

poxviruses, 140

Pratt, Mary Louise, 201–2

precariat, xii

preparedness, 137, 156–57, 167–68, 199

Preston, Richard, 134, 145, 154–55

primates: alienation attributed to, 102–3; domestication of, 130–31; Indian export ban, 112–13; free-ranging colonies, 114; horror film depictions of, 117; humanization of, 125; indoor captive institutions, 114; and myth of black male rapist, 121; national system for importation and breeding, 109–16; nuclear weapons research on, 112; promotion of term by National Institutes of Health, 125; symbol of occupation of Puerto Rico, 126–29; treatment in captivity, 115–16

primate trade, 20–21, 103, 110–11

Primate Visions (Haraway), xv

Project for a New American Century, 147–48, 166–67

Promin, 62

prophylaxis stations, 81, 87

prostitution. *See* sex work

protests: at Guantánamo, 180; at Krome immigration jail, 184; of treatment of Haitian refugees, 188; of US Panamá cooperative control program, 90–92

psychoanalysis, 16

Puar, Jasbir, xi, 69

Public Health Service, US, 71

public opinion, 150

public relations, 192

public/secret divide, 150–51, 153

Puerto Rico, 106–16, 126–29; assertion of control of Columbia School of Tropical Medicine, 111–12; monkeys as symbol of occupation in, 126–29; soldiers from, 83; suspicion of US government, 127–28

Punta Santiago, Puerto Rico, 110

putrid foods, 46

quarantines, 18–19, 21–23, 46; end of forced segregation for Hansen's disease, 62; escape from, 93; export of US system to Asia, 37; fugitives from, 54, 61; gendered, 81; of Hansen's disease patients, 31, 33–34; of HIV-positive people, 64, 171, 175; idealization of, 52; legal challenge to, 34; medical basis for, 184; militarized, 162–63, 171; necessity of, 67; outsides to, 61; penicillin treatment as justification for, 90; port, 3, 46; in service of common good, 36; of sex workers, 92; of venereal disease suspects, 73; violence of, 170; of women's bodies, 78

queerness, viii, 209–10n10; and desire, 53, 56; of disabled sexuality, 56; and kinship, 61–62; of life, 8, 195

Qur'an, 192

race: and AIDS response, 171; apes symbolic of, 121; Asia-Pacific racial categories, 49, 59; blackness, 51; and dependency, 51; disability associated with, 43, 58–59; disease associated with, 5–8, 13–14, 17, 22, 46, 64; and embodied memory of pathogens, 13; and epidemiology, 175; and fear of vivisection, 122; fears of, 136; Hansen's disease associated with, 49; and neoconservative policy, 147; in Panamá Canal Zone, 216n5; and paranoia, 151–52; profiling by, 6, 59, 74, 88; role in US security state, 197–98; segregation and, 72; and settler colonialism, 139–43; as sexual risk, 83–85; smallpox associated with,

163; and South Asians, 46; and uncertainty of social relations, 223n26; undertheorization of, 208n15; venereal disease associated with, 71–72, 78; whiteness threatened by, 81–83, 117–18, 124

Radford, Benjamin, 221n47

Rats, Lice, and History (Zinsser), 195

Reagan, Ronald, 125

Rebuilding America's Defenses (Project for a New American Century), 148

red-light districts, 77

reform discourses, 67

refoulement, 173, 177, 182, 188–89

refusal of treatment, 188

regeneration (anticolonial), 25, 212n40

Regional Primate Research Centers, 113, 125

"Reontologizing Race" (Saldanha), 209n8

"Report Relating to the Origin and Prevalence of Leprosy in the United States, A" (Wyman), 36, 38–48

reservoirs of disease, 72, 80

resistance: to colonization, 24–25, 54–55, 201; idealized, 58–59

resource extraction, 20–21

rhesus macaques, 101–4; availability of, 114; demand for, 108; domestication of, 116; escapes of, 126–29; harvesting of, 109; importation from India, 108–9; indoor captive breeding of, 116; moratorium on exports of, 112; native ecologies of, 110; smallpox experiments on, 145; use in polio research, 103, 107

Rice, Condoleezza, 166

risk: of AIDS, 172; catastrophic, 14–15, 167; diffusion of risk control, 7; displacements of, 20; intensification of inequality, 202; logic of systemic, 199; mapping of, 162; in military-civilian sex, 86; public feeling of, 17; racialization of, 6–8, 74, 198; representation of, 9; cost-benefit thinking and, 167; speculation, 23, 138–39, 144; surplus populations exposed to, xiv; and suspense in fiction, 155; of venereal disease, 72

Ritter, Scott, 157

Rockefeller Institute, 107

Rogers, Ginger, 125

rogue states, 148, 153

Roosevelt, Franklin D., 144

Rove, Karl, 148–50, 203

Roy, Parama, 30

Rubin, James, 149

Rumsfeld, Donald, 1–4, 6, 14, 134, 147, 154, 167, 200

Russia: bioweapons programs, 156; retention of smallpox stocks, 144

Sabin, Albert, 105

sacrifice, 7, 123–24, 145–46, 209n2

Saldanha, Arun, 6, 209n8

Salk, Jonas, 105, 109

Salvatore, Richard, 73

same-sex desire, 53, 56–58

sanitation, 72–75, 185–86; and disease transmission, 46; invasive techniques of, 75; corridors, 19, 74–75. *See also* hygiene

Santo Tomás Hospital, Panamá, 91

Scalia, Antonin, 203

scenario-based exercises, 158

Schelling, Thomas, 148

Schmidt, Leon, 114–15

Science (journal), 110

Seabrook, William, 212n4

secrecy, 135, 150–53

security, viii; and austerity, 199; and creation of uncertainty, 148; emergent forms of, 196; freedoms sacrificed for, 7; of laboratories, 131; neoconservative emphasis on, 147; and neoliberalism, 196–97; planning, 162; preemptive, 167; rhesus macaques as materials for, 114; technology as threat to, 125. *See also* biosecurity

security state, 23, 150–51

Senate Subcommittee on the Pacific Islands and Puerto Rico, 47

sensationalism, 214n37; of Hansen's disease, 47, 49, 54

sensitive but unclassified (SBU) information, 135, 150–51

Serlin, David, 105

T cell counts, 185
technology: biomedical, xii; history of, 207n5; restriction of knowledge, 150
territoriality and circulation, 204
terrorism: biological, 138; deterrence of, 148; and Islamic rogue trope, 139, 149
Time magazine, 112
Tito, Joseph, 161
Tomso, Gregory, 53–54
TOPOFF exercise, 158
traders, 54–55
transitions: bodily, 8–11; as degeneration or animalization, 47; of Hansen's disease, 31, 39
transit of empire, 143, 209n24
trans studies, 9
Tree of Life (Condé), 94
tropical landscape, idealized, 111
tropical medicine, 15, 19, 31, 50–52, 72–74, 109, 137
Trouillot, Michel-Rolph, 176
True Story of Kaluaikoʻolau, The (Kaluaikoʻolau), 61
tuberculosis, 33, 75, 106, 111, 175, 185
Turse, Nicholas, 66
Tuskegee Syphilis Study, 71
"Typhoid" Mary Mallon, 12, 20, 34

uncertainty, 148–49; affective generation of, 156, 158; engineered, 143–50; about Hansen's disease, 39; inevitability of, 147, 154; in international relations, 162; militarized response to, 161; racialization of, 152, 198; Saddam Hussein as figure of, 153
United Front for the Defense of the Lajas Valley, 128
United Nations, 156
United Nations Special Commission (UNSCOM), 146–47, 149, 157
United States: AIDS response, 23; bioweapons development, 22, 144, 152, 156; danger of Hansen's disease to, 40; delimitation of sovereignty of, 69–70, 76; democracy, vii; disavowal of support for Haitian dictatorships by, 177; exceptionalism, xv, 9, 18–19, 203; expansion into Pacific and Caribbean, 32–33; funding of international health movement, 21; influences on Panamá, 95; Iraq occupation, 165; plans to remove Saddam Hussein, 148; restriction of weapons technology, 150; retention of smallpox stocks, 144; settler expansion, 14; sponsorship of Iraq bioweapons program, 146; support for Iraq's unconventional weapons programs, 222n25
United States empire, 4–5, 23–24, 166, 203, 211n35
University of Puerto Rico, 111, 220n28
US Caribbean Command, Panama Canal Department, 80
US Strategic Command, 148

vaccination: experimental, 108; Hansen's disease attributed to, 46; objectors to, 2; politicizing of, 200; stockpiles, 167; universalization of, 7
Varela, Francisco, xv
Venereal Disease Circular (US Public Health Service), 82
Venereal Disease Hospital, Panamá City, 73
venereal diseases, 18–19, 71–99; case histories posted in barracks, 85–86; connected to Hansen's disease, 80; control of, 78, 89; detention for, 81; feminized representation of, 80; military report about, 83; visualization of, 79; white femininity threatened by, 81–83
"Venereal Disease Situation in the Panama Canal Zone, The" (Wenger), 71
Verdieu, Elma, 187
Versuchpersonen (human guinea pigs), 65–66
vice, 72–73, 78, 81, 85, 93–99
Vietnam War refugee crisis, 226n18
violence: interspecies, 26; of medical technology, 119; psychic and ecological, 98
vital fix (Marxism), xiv
vitalist thought, 16